Books are to be returned on or bef
the last date below.

24 DEC 1991

ILSI Human Nutrition Reviews

Series Editor: Ian Macdonald

Already published:

Sweetness
Edited by John Dobbing

Calcium in Human Biology
Edited by B. E. C. Nordin

Sucrose: Nutritional and Safety Aspects
Edited by Gaston Vettorazzi and Ian Macdonald

Zinc in Human Biology
Edited by C. F. Mills

Dietary Starches and Sugars in Man: A Comparison
Edited by John Dobbing

Forthcoming titles in the series:

Modern Lifestyles and Low Energy Intake
Edited by K. Pietrzik

Thirst: Physiological and Psychological Aspects
Edited by D. A. Booth and D. J. Ramsay

Diet and Behavior:
Multidisciplinary
Approaches

G. Harvey Anderson

Editor-in-Chief

Norman A. Krasnegor • Gregory D. Miller
Artemis P. Simopoulos (Editors)

With 26 Figures

Springer-Verlag
London Berlin Heidelberg New York
Paris Tokyo Hong Kong

Editor-in-Chief
G. Harvey Anderson, PhD
Department of Nutritional Science, University of Toronto

Editors
Norman A. Krasnegor, PhD
Gregory D. Miller, PhD
Artemis P. Simopoulos, MD
(Addresses on Contributor list: xv–xvii)

Series Editor
Ian Macdonald, MD, PhD, DSc, FIBiol
ILSI Nutrition Co-ordinator, Professor of Applied Physiology,
Emeritus, Guy's Hospital Medical and Dental School,
St Thomas's Street, London, SE1 9RT, UK

ISBN 3–540–19595–5 Springer-Verlag Berlin Heidelberg New York
ISBN 0–387–19595–5 Springer-Verlag New York Berlin Heidelberg
ISSN 0–936–4072

British Library Cataloguing in Publication Data
Anderson, G. Harvey, *1941–*
Diet and behavior.
1. Man. Behavior. Effects of nutrition   I. Title   155.91
ISBN 3–540–19595–5

Library of Congress Cataloging-in-Publication Data
Diet and behavior: multidiciplinary approaches/edited by G. Harvey Anderson
p. cm.   Papers from a workshop sponsored by the Diet and Behavior Committee of
the International Life Sciences Institute-Nutrition Foundation, Washington, D.C.
ISBN 0–387–19595–5 (U.S.)
1. Nutrition–Psychological aspects–Congresses. 2. Human behavior–Nutritional
aspects–Congresses.   3. Food preferences–Congresses.
I. Anderson, G. Harvey, 1941–   . II. International Life Sciences Institute-Nutrition
Foundation. Diet and Behavior Committee.
QP141.A1D47 1990   612′.3′019—dc20   90–9541
                                                                          CIP

Typeset by Wilmaset, Birkenhead, Wirral
Printed by Alden Press Ltd, Osney Mead, Oxford

2128/3916–543210 Printed on acid-free paper

Foreword

The topic of Diet and Behavior is one that has captured the public's attention and in the process has generated a considerable body of so-called facts based largely on "hearsay" evidence. Nevertheless the relationship between diet and behavior is very important. Scientific understanding and assessment are beginning to clarify the issues, but there is obviously much more to be learnt. This volume emphasizes that the scientific approach must be followed in studies of the relationship between diet and behavior and that the methods used should be those that are acceptable to scientists in both the behavioral and biological disciplines. Each of the chapters in this volume has been peer reviewed, in keeping with comparable volumes in this Series, and thus, of course, gives considerable credence and acceptability to the contents.

The interrelationships of diet and behavior affect each one of us as individuals in ways we may not yet appreciate, let alone understand. From a broader point of view, those concerned with public health issues will be interested in this relationship, whether in the handling of, say, children with behavioral problems or adults with obesity, and likewise diet and behavior is a subject that is of considerable interest to the food industry. It is therefore essential that the understanding of diet and behavior is based on so-called hard science such that advice offered and acted upon is more than a mere guess, hunch or "common sense".

The publication of *Diet and Behavior: Multidisciplinary Approaches* was made possible by the support of the International Life Sciences Institute – Nutrition Foundation (ILSI–NF). ILSI is a non-profit scientific foundation established to encourage and support research and educational programmes in nutrition, toxicology and food safety and to encourage cooperation in these programmes among scientists from universities, industry and government to facilitate the resolution of health and safety issues world-wide.

London
October 1989

Ian Macdonald
Series Editor

Preface

In recent years the notion that foods and nutrients influence brain function and behavior has generated widespread interest in the scientific community. However, scientists who have ventured into the area have often discovered controversy, ambiguous data and poor scientific methods.

For many scientists the area gained legitimacy during the 1970s with the discovery that the developed central nervous system was directly responsive to changing nutrient environments as reflected in the bloodstream. Thus, for example, it was shown that brain neurochemistry could be readily altered by the composition of food consumed. Whether or not food-induced changes in brain neurochemistry affect behavior is still a subject of controversy. However, a conceptual framework for examining some of the diet behavior linkages was provided, and proved useful. For example, the biochemical evidence that carbohydrates increase brain serotonin, an inhibitory neurotransmitter, proved useful in disproving the prevalently held view that sugar consumption caused hyperactivity in children. As predicted from the biochemistry, sugar loads tend to decrease rather than increase activity.

Many of the nutrition and behavior connections have not proven to be as easily resolved as that for sugar and hyperactivity of children. There are many reasons for this, but they include poor definitions of the behavioral phenomena and of diet composition, inadequate measurements of behaviors and of diet, and the uncertainty of cause and effect. Investigations in the area therefore, benefit greatly from multidisciplinary inputs into their design and in the analysis and interpretation of data.

Because the advancement of knowledge of the links between diet and behavior is dependent upon the use of appropriate methods, rigorous study design, and careful interpretation of data based on both the nutritional and behaviorial sciences, a workshop was held with the focus on methodologies. The workshop organizers hoped that, by challenging the participants representing several disciplines to examine methodologies, the development of research methods and standards would be advanced and that multidisciplinary interaction in research in the area would be recognized as essential.

To achieve the objectives, the workshop was organized into four sessions. In Session I, "Origins of Food Preference", the question of "why do individuals and population groups eat what they do?" was explored. Participants in Session II, "Measuring Behavioral Response", discussed the challenges and new opportunities for identifying, measuring, and operationalizing the subtle effects on behavior of diet and nutrients. Because of the interest in the application of epidemiology to nutritional diseases, its potential to assist in identifying diet and behavioral relationships was explored in Session III, "Epidemiological Studies". Finally, to highlight the need for multidisciplinary interaction, Session IV dealt with the "Integration of Research Methods in Diet and Behavior".

The publication arising from the presentations and discussions is meant to be of benefit both to those contemplating developing a research program and to experienced researchers in the area.

The workshop and its publication were made possible through the support of the Diet and Behavior Committee of the International Life Sciences Institute – Nutrition Foundation, Washington, DC.

Toronto   G. Harvey Anderson
October 1989   Editor-in-Chief

Contents

Contributors

G. H. Anderson, PhD
Department of Nutritional Science, University of Toronto,
150 College Street, Toronto, Ontario
M5S 1A8, Canada

G. K. Beauchamp, PhD
Monell Chemical Senses Center, 3500 Market Street, Philadelphia,
PA 19104, USA

T. W. Castonguay, PhD
Department of Human Nutrition and Food Systems, University of
Maryland, Marie Mount Hall, Room 3304, College Park,
MD 20742, USA

L. S. Crnic, PhD
Departments of Pediatrics and Psychiatry, C233, University of
Colorado School of Medicine, 4200 E. 9th Avenue, Denver,
CO 80262, USA

P. B. Dews, MB, ChB, PhD
Laboratory of Psychobiology, Harvard Medical School,
220 Longwood Avenue, Boston, MA 02115, USA

C. W. Gershenson, PhD
Institute for Disabilities Studies and Department of Psychology,
University of Minnesota, 2221 University Avenue, SE, Suite 145,
Minneapolis, MN 55414, USA

C. E. Greenwood, PhD
Department of Nutritional Science, University of Toronto,
150 College Street, Toronto, Ontario M5S 1A8, Canada

M. R. C. Greenwood, PhD
Dean of Graduate Studies and Professor of Nutrition, Department
of Nutrition, University of California–Davis, Davis,
CA 10128, USA

H. A. Guthrie, PhD
Department of Nutrition, The Pennsylvania State University,
126 Henderson Building South, University Park, PA 16802, USA

A. E. Harper, PhD
Departments of Nutritional Sciences and Biochemistry, College of
Agricultural and Life Sciences, University of Wisconsin–Madison,
Madison, WI 53706, USA

M. Hetherington, PhD
Biomedical Psychiatry, National Institute of Mental Health,
Building 10, Room 35–231, Bethesda, MD 20892, USA

T. Johns, PhD
School of Dietetics and Human Nutrition, MacDonald College of
McGill University, 21111 Lakeshore Road, Ste. Anne de Bellevue,
Quebec H9X 1C0, Canada

M. Kiely, PhD
New York State Institute for Basic Research in Developmental
Disabilities, 1050 Forest Hill Road, Staten Island, NY 10314, USA

N. A. Krasnegor, PhD
National Institutes of Health, National Institute of Child Health
and Human Development, CRMC, HLB, 9000 Rockville Pike,
EPN, Room 633, Bethesda, MD 20892, USA

H. V. Kuhnlein, PhD
School of Dietetics and Human Nutrition, MacDonald College of
McGill University, 21111 Lakeshore Road, Ste. Anne de Bellevue,
Quebec H9X 1C0, Canada

M. Krondl, PhD
Department of Nutritional Science, University of Toronto, 150
College Street, Toronto, Ontario M5S 1A8, Canada

G. D. Miller, PhD
Kraft General Foods, 801 Waukegan Road, Glenview, IL 60025,
USA

C. Pollock
Institute for Disabilities Studies and Department of Psychology,
University of Minnesota, 2221 University Avenue, SE, Suite 145,
Minneapolis, MN 55414, USA

F. A. Rhoads, MD
General Pediatric Ambulatory Center, Children's Hospital National
Medical Center, 111 Michigan Avenue, NW, Washington,
DC 20010, USA

G. G. Rhoads, MD, MPH
Division Director, Public Health and Medical Humanities,
University of Medicine and Dentistry of New Jersey, Robert Wood
Johnson Medical School, Piscataway, NJ 0885–5635, USA

J. Rodin, PhD
Department of Psychology, Yale University, 2 Hillstone Avenue,
PO Box 11A Yale Station, New Haven, CT 06520, USA

B. J. Rolls, PhD
Department of Psychiatry and Behavioral Sciences, The Johns
Hopkins University School of Medicine, Adolf Meyer Building,
600 North Wolfe Street, Baltimore, MD 21205, USA

H. J. Schmidt, PhD
New Jersey Medical School, 185 E. Orange Avenue, Newark, NJ
07103, USA

P. H. Shiono
National Institutes of Health, National Institute of Child Health
and Human Development, Prevention Research Program,
9000 Rockville Pike, Bethesda, MD 20892, USA

A. P. Simopoulos, MD
The Center for Genetics, Nutrition and Health, American
Association for World Health, 2001 S Street, NW, Suite 530,
Washington, DC 20009, USA

J. S. Stern, ScD
Department of Nutrition, Food Intake Laboratory, University of
California–Davis, Davis, CA 95616, USA

T. Thompson, PhD
Institute of Disabilities Studies and Department of Psychology,
University of Minnesota, 2221 University Avenue, SE, Suite 145,
Minneapolis, MN 55414, USA

P. M. Vietze, PhD
New York State Institute for Basic Research in Developmental
Disabilities, 1050 Forest Hill Road, Staten Island, NY 10314, USA

R. H. Ward, PhD
Department of Human Genetics, The University of Utah, School of
Medicine, 501 Wintrobe Building, Salt Lake City, UT 84132, USA

Section 1
Origins of Food Preference

Introduction

G. H. Anderson

Associations found between diet and behavior often do not provide evidence for cause and effect unless specific hypotheses are formulated and experimental testing occurs (Anderson and Hrboticky 1986). However, it is obvious that the relationship can go either way: behavior can determine what foods and nutrients are consumed and, conversely, foods and nutrients can influence behavior. In this first section the multitude of factors involved, including cultural, psychological, and behavioral determinants of food selection, are examined. These factors are difficult to define and their relative importance is not necessarily constant.

The first chapter by Krondl examines four conceptual models used by social scientists to provide a framework for understanding food selection behavior. She discusses the sensory, belief, ecological, and perceptual models, the increasing complexity of which is caused by both the numbers of, and interactions between, factors influencing food selection. She concludes that none of the models provides a satisfactory explanation of food choice behavior and that a further evaluation of the separate components of the systems and their interrelationships is required.

Johns and Kuhnlein describe in the second chapter how food selection and behavior are influenced by culture. They point to the difficulty in defining and measuring culture, but provide examples of how cultural patterns of food choice within human populations are consistent with basic biological needs. It is also suggested that the methodologic strengths of cultural anthropology require a better understanding from biologists.

The third chapter, "Biological Determinants of Food Preferences in Humans" by Schmidt and Beauchamp, suggests that putative brain mechanisms based on metabolic and sensory inputs might contribute to our selection of a balanced diet and avoidance of toxic substances. It is concluded, however, that at present we have little understanding of the biological factors underlying food choice and preference, and the extent to which biological factors explain human eating behavior.

In her discussion of this section Guthrie observes that individual scientists often appear to be narrowly focused and based in their own discipline and perspective. When the effort is made, however, it becomes clear that there is more common ground between scientists of differing backgrounds than was first evident.

Reference

Anderson GH, Hrboticky N (1986) Approaches to assessing the dietary component of the diet–behavior connection. Nutr Rev 44:42–50

Chapter 1

Conceptual Models

M. Krondl

Introduction

The answer to the seemingly simple question "why do individuals and population groups eat what they do?" lies within a complex decision-making process where the decision maker responds to a food stimulus by selecting or rejecting the food. This choice is a behavioral act (Bodenstedt 1983) and it occurs, independently of the actual ingestion of the food, during the pre-ingestive phase of eating (Randall 1982). The difficulty in answering the initial question arises because the choice of foods is food or food component specific and involves a multitude of factors.

Conceptual models of food selection provide a structure within which researchers can examine the role of the factors influencing food choice behavior and can develop an understanding of the mechanisms involved. The models are constantly developing because food choice behavior is a dynamic process. As society changes and becomes more complex, a change occurs in the relationship between food resources and a person's general food behavior (Quick 1970). For example, when consumers contribute little of their own energy to the creation of food supplies, the biological relationships between physical output, energy needs and food resources are lost (Krondl and Boxen 1975). These relationships were probably better balanced among hunter-gatherers than among consumers in affluent urban societies of today where, for example, foods which are altered by production techniques take on new meanings. New types of foods and food varieties may reflect little of their original form and composition; thus fruit flavored crystals have little in common with fresh orange juice. In addition, modern technology can affect the processes leading to food choices, particularly the sensory-specific and the belief–attitude components.

Conceptual models may vary according to the number of factors under investigation and the interrelationships between them. Thus a study of food preference, as measured by the liking of one food or food component more than another, can be based on a simpler model than investigations of attitudes to foods. The latter encompass beliefs and perceptions of foods that originate within and/or outside the individual.

The purpose of this chapter is to review four conceptual models developed to provide a framework for understanding food selection behavior. They are presented in ascending order of complexity. For ease in identification they are termed sensory, belief, ecological and perceptual. In the following discussion each basic model will be illustrated by means of specific studies and assessed in terms of its relative merits.

Sensory Model

The sensory model is designed to measure a person's reaction to the components of food stimuli. It is the most simple of the conceptual models of food choice. It considers a limited number of components, such as one flavor attribute of a food. This attribute may be assessed by quantitative measure of intensity or degree of liking. An example of this approach is described in a study which compared adolescents and adults (Desor et al. 1977). Each subject was presented with four cups containing different concentrations of sucrose, asked to taste the samples without swallowing and then to rank the concentrations in order of preference. The order in which the concentrations were sampled varied across subjects. More of the younger subjects (approximately 50%) selected higher concentrations of sucrose as their most preferred. The adults were approximately evenly distributed in terms of preference among all four sweetness levels.

The sensory model is most appropriately examined in a laboratory setting where specific factors operating under controlled conditions can be identified. It places emphasis on the relationship between the stimulus and food behavior with little regard for the characteristics of the person. Studies using this model are not intended to provide an explanation for the mechanisms involved in food preference.

There are limitations inherent in the simplicity of the sensory model. These include subject individuality in taste acuity, experience with the specific taste and degree of satiation in terms of liking or disliking. Because the range of intensity within which stimuli arouse pleasure is different for different people (Cabanac 1979), a narrow range of concentrations of the stimulus may fail to provoke a response. In addition the test medium may not be similar to foods encountered in the real world and this will introduce contextual artificiality into the study (Pangborn 1970).

Responses to sensory stimuli are highly individualistic among subjects due to biological and other factors (Pangborn 1981). Biological factors may account for certain fundamental taste preferences since innate preferences for sweet tastes and aversion to bitter ones have been demonstrated (Steiner 1973). Innate responses to stimuli have obvious survival value for the omnivore. A safe and quick source of energy is associated with sweet foods whereas harmful substances are associated with bitter tastes (Rozin 1976).

Generalizations about preferences caused by sensations produced by food substances must be made with caution, however, since experience of the taste sensation may increase its attractiveness. Humans may come to accept and like a number of innately unpalatable foods such as, for example, chili pepper (Rozin and Fallon 1981). A further problem arises in the application of the sensory model as most foods are eaten in combination and the taste components of each food interact. The outcome is enhancement of sensation from one component and masking of another. The best known example is the use of sweeteners in bitter tasting beverages such as coffee where the food–drink combination masks the bitter reaction with a sweet sensation.

Food preference has to relate to the basic food function which is to ensure a supply for the body of the important nutrients and to protect the body from harmful substances. The relationship between food and the response to it is specific at a given level of metabolic permissiveness which designates the palatability of a particular food at the time of consumption (Le Magnen 1985). Similarly Cabanac (1971) suggested that pleasure or displeasure of a sensation is not stimulus-bound but depends entirely on physiologic signals related to usefulness of the stimulus in relation to the need of the subject. He proposed the term "alliesthesia" to describe a change in sensation of pleasantness depending on the hunger–satiety state. The sensory properties of foods have a major influence on the hedonic changes occurring during and after eating a meal (Rolls et al. 1986). Thus, in studying sensory responses, factors such as the time of day and food intake at the previous meal must be closely controlled. Sensory models recognize that sensory aspects of foods may arise from oral sensations complemented by gastrointestinal and cephalic responses.

Food selection can be predicted, based on sensory aspects of foods, which makes this information useful in areas of food marketing and product development. For example, the sensory model was applied in studies of food preferences among 3885 members of the US Armed Forces (Meiselman and Waterman 1978). The respondents were queried about food likes and dislikes and asked how frequently they would like to eat each of 378 food items. For analysis the foods were grouped into 33 food classes, primarily based on their use in military menus. Items within each class were ranked according to the mean values for degree of liking or disliking of the food and frequency of wanting the food served (Table 1.1; Wyant and Meiselman 1984). The women preferred vegetables, salads and fruit while men preferred meat products. Blacks preferred fruit and fruit juices more than did whites. The findings are in accordance with the results of studies measuring foods actually eaten.

In summary, the sensory approach is useful in measuring reactions to the components of food stimuli, as well as in measuring other responses, such as hunger for specific nutrients. The sensory model facilitates the development of an understanding of the link between food and biological determinants of food

Table 1.1. Male/female differences in food liking among members of the US armed forces

Liked better by men	Liked better by women
Beer	Appetizers
Eggs	Potato
Breakfast meats	Green vegetables
Meats	Other vegetables
Stews	Vegetable salads
Short order sandwiches	Tossed green salads
Pies	Fresh fruit

From Wyant and Meiselman (1984).

acceptance; it is not, however, designed to identify the individual's attitude to food. This domain of the human response to food stimuli is illustrated by the belief model to be described next.

Belief Model

The belief model is designed to study attitudes affecting behavior. A health belief model is used in prediction of food selection. In an environment with an abundant food supply where people are guided more by sensory than by hunger cues, there may be unhealthful food choices. In fact the diseases of dietary excess and imbalance now rank among the leading causes of illness and death in North America (The Surgeon General's Report 1988). The health belief model of food choice was first proposed by Fishbein (Fishbein and Ajzen 1975). This model was welcomed by nutrition educators because its application allows one to predict the impact of an intervention aimed at changing food choices. The model was founded on the theory that an individual's intention to perform some behavior is its best predictor (Ajzen and Fishbein 1980). In the process of assessing this intention, an estimate is first derived of individuals' beliefs concerning their behavior, such as the choice of a specific food. Secondly, the belief of appropriateness of that behavior by people important to the individual is assessed. The values of the latter are considered to be a reference for the judgment of the experimental subjects or the subjective norm. Both these factors, the individual's attitude and the subjective norm, are weighted according to the relative importance to the subjects of both components in determining their intention for exhibiting the specific behavior (Fig. 1.1).

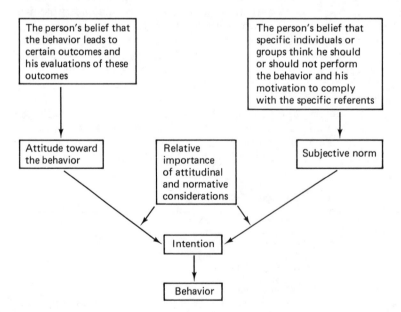

Fig. 1.1. Belief model (Ajzen and Fishbein 1980). Reprinted by permission of Prentice-Hall, Englewood Cliffs.

The application of the Fishbein model is illustrated in a study of prediction of eating at a fast-food restaurant (Axelson et al. 1983). Examples of statements used for the measurements of salient beliefs and the individual's evaluation are as follows:

1. Eating at a fast-food hamburger restaurant in the next two weeks is:
   good————bad

2. Having an inexpensive meal is:
   good————bad

Examples of statements used to assess the subjective norm and the individual's motivation to comply with the wishes of those influential people are as follows:

1. Most people who are important to me think that I should eat at a fast-food hamburger restaurant in the next two weeks:
   likely————unlikely

2. Generally speaking, I want to do or I want not to do what I think my parents think that I should do:
   want to do————want not to do

Other studies, such as those concerning attitudes to eating foods high in fat, adding table salt to foods and eating snack food (Shepherd and Stockley 1985), further illustrate the application of the Fishbein model. In all of these studies, the attitude to the behavior was a better predictor of behavioral intention than was the subjective norm. Another study designed using the Fishbein model was related to consumer responses to low salt bread (Tuorila-Ollikainen et al. 1986). The model was extended to include the degree of liking/disliking of the products, which greatly improved the predictive power because the sensory factor was stronger than the belief about healthfulness of the food in determining the intention to make a food choice.

In summary, although strong in terms of health implications, the model does not measure the impact of the environment on food choices and does not provide an understanding of the mechanisms involved.

Ecologic Model

Ecologic models of food choice center individuals within their physical and social environment. They complement the previously discussed sensory and belief models. The model (Fig. 1.2) used by nutritional anthropologists (Jerome et al. 1980) indicates the interaction of the many factors affecting persons in their food choice behavior. The physical environment establishes the conditions for food production. Technology assists in food production and distribution. Both factors influence availability of foods and the potential for their acquisition. The many economic and political structures, parts of the social organization, affect the access to foods mainly by their impact on food price and the available food

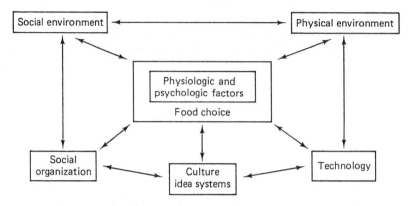

Fig. 1.2. Ecologic model (after Jerome et al. 1980).

dollar which influences purchasing power. Foods such as bread, flour, potatoes, sugar and jams, which are viewed as staple items, hold a static position regardless of income and price fluctuations. In contrast, significant increases in consumption of dairy, meat, fish, fruit and vegetable products occur when income increases. A reverse trend occurs for condensed milk and margarine (McKenzie 1979). When purchasing power is limited, access to food is affected and sensory and health motives become secondary considerations in food choices.

The social environment is primarily responsible for the education of individuals and may affect their belief about healthfulness of foods. Cultural and ideologic systems include ideas about the health role of food and religious beliefs involving foods, as illustrated by the following two examples. First, Moore (1970) assessed food choice in light of the social role of food and food taboos in 383 non-industrial societies. On one hand societies demonstrated a preference for meat, whenever available, due to its central role in social or ceremonial meals; on the other hand some types of meat, especially pork, were the subject of many prohibitive religious laws, practiced by Moslems among others. Rice was similar to meat in being highly desired but in short supply for much of the population in the study. A second example of an anthropologic nature is the study by Simoons (1973) who investigated reasons for non-milking practices in areas of Africa and Asia. Geography, climate and existing food practices did not favour familiarization with milk, leading to its lack of acceptance.

Both of the above examples suggest that the physical environment is likely to be highly influential in traditional societies; however, food choice in modern societies is highly influenced by technology. This can be illustrated by several examples, one of which is the production of convenience food. Convenience in foods is generally defined as anything that saves or simplifies work and adds to one's ease or comfort. More specifically, convenience foods are those which have service added to the basic ingredients to reduce the amount of preparation required in the home (Glicksman 1971). A second example of the impact of technology is in changes in the composition of foods in response to consumers' health concerns. For example, the evidence of the relationship between high fat consumption and coronary heart disease led to the demand for leaner meat, resulting in changes in beef production practices (Wood et al. 1988). Another type of technologic advance, evident in information transfer, developed from the conventional forms of printed advertising and radio broadcasting to two-way television available through cable systems and direct broadcasting by satellite. Thus its application in food marketing and health promotion is likely to increase and to be substantial in its impact on food choices.

In summary, this model emphasizes the environmental impact on food choices and considers the person making them but does not systematically explore the range of food meanings and their influence on individual food behavior.

Food Perception Model

The food perception model of food selection behavior incorporates the main components of the sensory, belief and ecological models. It can be viewed as three arms addressing the questions "why?", "who?" and "where?" and linking them to the central issue of food choice (Fig. 1.3). The arm designated as "why?" refers to food perceptions which include components found in the sensory and belief models. The second arm, representing the question "who?", identifies persons in terms of their specific biological needs such as those due to heredity, gender, age, health and activity. In addition, psychologic individuality is addressed; this may describe the target population in terms of personal state of mind such as mental depression which influences food selection (Letsov and Price 1987). The third arm of the food perception model, asking the question "where?", relates to the physical and social environments within which the food

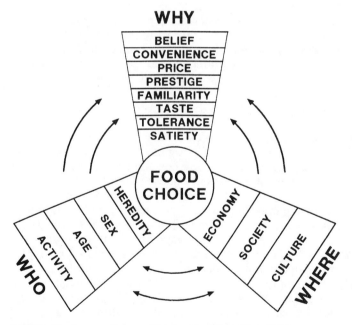

Fig. 1.3. Food perception model (after Krondl and Lau 1979).

choice occurs. The physical environment refers to the place and time of food choice, while the social environment encompasses the social and cultural norms that influence the individual's relation to foods. In addition their economic status belongs within the social category of variables. All the personal and environmental factors were included in the ecological model.

The "food perception model" was inspired by the work of Pilgrim (1957). In his early model Pilgrim defined the term food perception as the outcome of sensations from food stimuli, modified by a person's physiology and attitudes. While Pilgrim emphasized that the person's perception varied according to his or her state of hunger, he did not categorize the individual perceptions in terms of their development. It was Fewster and colleagues (1973) who provided this dimension and method of measurement.

The concept of food perceptions is best described by Olson (1981) and Barker (1982). Food perceptions can be viewed as the outcomes of previous real or vicarious food experiences. They form structures stored as memory schemata, which may be activated on new encounters with foods, to provide an evaluation within a given situation. This evaluation leads to the formulation of a judgment about a food and results in acceptance or rejection. Thus, the whole process is a food-oriented interpretation of information processing involving memory. Well-developed, frequently used schemata may appear to operate automatically. Such well-learned schemata allow the rapid judgments and evaluations typical of our everyday life.

Perceptions form an imaginary barrier between food availability and food choice. Krondl and Lau (1978) have delineated seven categories of perceptions within this barrier: satiety, tolerance and taste as found in the sensory model, and healthfulness from the belief model; price, convenience and prestige of foods are additional categories. Familiarity is included with the perceptions in order to assess length of previous exposures to the food, a factor which also affects food choice. Satiety is an important category because hunger triggers yearning for satisfaction and contentment. Tolerance, or conversely intolerance, relates to any ill effect experienced in an encounter with a food that will cause its rejection. Taste refers to sensory response. Healthfulness has a social and cultural connotation. Price refers to the subjective evaluation of the cost of food. Convenience depends on the ease with which a food may be prepared by the choice-maker. Prestige describes the perceived suitability of a food for important guests or special occasions.

For measurement of the perceptions, scales similar to those suggested by Fewster (Fewster et al. 1973) can be used. The significance of individual food perceptions in food choices was established by correlating the rating of the individual perception of a food with its frequency of use (Krondl and Lau 1982). An example illustrating the use of this model (Fig. 1.4) was a study of food perceptions and their relationship to the use of eight foods by male and female young and elderly persons (Krondl and Coleman 1988). The foods were whole wheat bread, beef liver, frozen fish fillet, 2% milk, margarine, lettuce, squash and apple juice. Unpleasant tastes such as those found to be associated with beef liver, fish or squash were related to food rejection. Perceived healthfulness of the foods ranked second to taste for elderly persons but fourth for adolescents as an associative factor in food use on the strengths of the simple correlations. Perceptions of tolerance and satiety ranked second and third as factors of significance in relation to use of selected foods among the young. The other

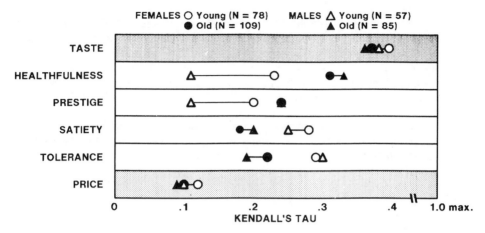

Fig. 1.4. Correlations between food use and food perceptions for young and elderly people for eight foods: whole wheat bread, beef liver, frozen fish fillet, 2% milk, margarine, lettuce, squash, apple juice (Krondl and Coleman 1988).

perception categories were found to be less important in food choices in this study.

Another application of the food perception model illustrates its value in assessing the dynamics of the food selection process. A comparison of first- and second-generation immigrant Chinese adolescents with Anglo-Canadian peers (Krondl et al. 1984) showed changes in perception of taste of foods in terms of the length of time spent in Canada. The response of the first generation differed from that of the second generation. The latter resembled more closely the response of the Anglo-Canadians, the dominant culture. The pooled responses of the Chinese sample showed a similar, but even stronger impact of acculturation in the change in perception of healthfulness. The second-generation Chinese were even closer to Anglo-Canadians in their perception of food healthfulness than in perceived taste.

In summary, the food perception model considers the relationships between the environment, the person, the psychological evaluation of foods by the person, and their outcome – the choice of foods. Thus its encompasses components of the three previously described models. The experiments emphasized the measurement of the strength of food perceptions in relation to the frequency of food use and controlled for the environmental and personal factors.

Conceptual Models

Conceptual models identify key factors and suggest their pathway. Thus they simplify a complex reality such as food selection and become indispensable in designing research in this area. In this chapter individual factors of food selection have been incorporated stepwise into the pathway leading from food stimulus to a person's behavioral response. The discussion of the models suggested that both the simple approach in the study of interaction between a limited number of factors affecting food selection, as well as the study of the concert of many factors, should be carried out separately and simultaneously. On account of both the physiologic and psychologic nature of the factors, better understanding of conceptualization and methodologies used by basic and social scientists has to be developed and closer cooperation established between them in the study of food selection behavior.

Conceptual models will become even more important in the future than they are now in light of the anticipated disturbing reality of a society that is undergoing changes at a speed unprecedented in human history.

Summary

This review of the sensory, belief, ecological and perceptual models of food selection behavior has shown how each of them, in sequence, contains additional components in an effort to deal with the complexity of mechanisms involving food choices. It is to be expected that while individual models will continue to be used in special lines of research, the awareness of other approaches will allow for the inclusion of components which will add to the value of the research and to the applicability of the results. As shown in the example used in connection with the belief model, the addition of the sensory component added to its strength. The increasing complexity of the social environment, its reciprocal impact on the physical environment and on individuals is already evident from the perceived healthfulness of low-fat foods outweighing in food choices the liking for fatty foods as the result of trends in health promotion.

The question of why people eat as they do can be fully answered only when all the separate components of the complicated system of food choices and their interrelationships are understood. This will require a continued evolution of the conceptual framework for the models of food choice behavior. Further, there has to occur elaboration of individual components and clearer delineation of the pathway of their interaction.

References

Ajzen I, Fishbein M (1980) Understanding attitudes and predicting social behavior. Prentice-Hall, Englewood Cliffs

Axelson ML, Brinberg D, Durand JH (1983) Eating at a fast-food restaurant. A social psychological analysis. J Nutr Educ 15:94–98

Barker LM (1982) Building memories for foods. In: Barker LM (ed) The psychology of food selection. AVI, Westport, pp 85–100

Bodenstedt, AA (1983) Development and present state of social science research in nutritional behaviour in the Federal Republic of Germany. Ernaehrungs-Umschau [Suppl] 30:9–13

Cabanac M (1971) Psychological role of pleasure. Science 173:1103–1107

Cabanac M (1979) Sensory pleasure. Q Rev Biol 54(1):1

Desor JA, Moller D, Green LS (1977) Preference for sweet in humans: infants, children and adults. In: Weiffenback JM (ed) Taste and development. US Department of HEW, Bethesda, pp 161–172

Fewster J, Bostian L, Powers R (1973) Measuring the connotative meanings of foods. Home Econ Res J 2:44–45

Fishbein M, Ajzen I (1975) Belief, attitude, intention and behavior: an introduction to theory and research. Addison-Wesley, Reading, Mass

Glicksman M (1971) Fabricated foods. In: CRC critical reviews in food techonology, vol 2. CRC Press, Cleveland

Jerome NW, Pelto GH, Kandel RF (1980) An ecological approach to nutritional anthropology. In: Jerome NW, Kandel RF, Pelto GH (eds) Nutritional anthropology. Redgrave, New York, pp 13–45

Krondl M, Boxen G (1975) Nutritional behavior, food resources and energy. In: Arnott ML (ed) Gastronomy. The anthropology of food and food habits. Mouton, The Hague, pp 113–120

Krondl M, Coleman P (1988) The role of food perceptions in food use. In: Winick M (ed) Control of appetite. Wiley, New York, pp 53–78

Krondl M, Lau D (1978) Food habit modification as a public health measure. Can J Public Health 69(1):99–103

Krondl M, Lau D (1982) Social determinants in human food selection. In: Barker LM (ed) The psychobiology of human food selection. AVI, Westport, pp 139–152

Krondl M, Hrboticky N, Coleman P (1984) Adapting to cultural changes in food habits. In: White PL, Selvey N (eds) Malnutrition: determinants and consequence. Alan R Liss, New York, pp 221–229

Le Magnen J (1985) Hunger. Cambridge University Press, Cambridge

Letsov AP, Price LS (1987) Health, aging and nutrition: an overview. Clin Geriatr Med 3:253

McKenzie JC (1979) Economic influences on food choices. In: Turner M (ed) Nutrition and lifestyles. Applied Science Publishers, London, pp 91–103

Meiselman HL, Waterman D (1978) Food preferences of enlisted personnel in the Armed Forces. J Am Diet Assoc 73:621–629

Moore FW (1970) Food habits in non-industrial societies. In: Dupont J (ed) Dimensions of nutrition. Colorado Associated University Press, Colorado, pp 133–135

Olson J (1981) The importance of cognitive processes and existing knowledge structures for understanding food acceptance. In: Solms J, Hall RL (eds) Criteria of food acceptance. Foster Verlag, Zurich, pp 69–81

Pangborn RM (1970) Individual variation in affective responses to taste stimuli. Science 21:125–126

Pangborn RM (1981) Individuality in responses to sensory stimuli. In: Solms J, Hall RL (eds) Criteria of food acceptance. Foster Verlag, Zurich, pp 177–219

Pilgrim F (1957) The components of food acceptance and their measurement. Am J Clin Nutr 5:171–175

Quick HF (1970) Geographic dimensions of human ecology. In: Dupont J (ed) Dimensions of nutrition. Colorado Associated University Press, Colorado, pp 133–153

Randall E (1982) Food preferences as a determinant of food behavior. In: Sanjur D (ed) Social and cultural perspectives in nutrition. Prentice-Hall, Englewood Cliffs, pp 123–145

Rolls BJ, Hetherington M, Burley VJ, van Duijenvoorde PM (1986) Changing hedonic responses to foods during and after a meal. In: Kare MR and Brend JG (eds) Interaction of the chemical senses with nutrition. Academic Press, New York, pp 247–268

Rozin P (1976) The selection of foods by rats, humans and other animals. In: Rosenblatt J, Hinde RA, Beer C, Shaw E (eds) Advances in the study of behavior. Academic Press, New York, pp 21–76

Rozin P, Fallon AE (1981) The acquisition of likes and dislikes for foods. In: Solms J, Hall RL (eds) Criteria of food acceptance. Foster Verlag, Zurich, pp 35–47

Shepherd R, Stockley L (1985) Fat consumption and attitudes towards food with a high fat content. Hum Nutr Appl Nutr 39A:431–442

Simoons FJ (1973) The determinants of dairying and milk use in the old world: ecological, physiological and cultural. Ecol Food Nutr 2:83–90

Steiner JE (1973) The human gustofacial response. In: Bosens JF (ed) Fourth symposium on oral sensation and perception. Development of the fetus and infant. US Government Press Office, Washington, DC, pp 254–278

The Surgeon General's report on nutrition and health (1988). US Department of Health and Human Services, Public Health Service, DHHS (PHS) Publication No. 88–50210, Washington, DC

Tuorila-Ollikainen H, Lahteenmaki L, Salovaara H (1986) Attitudes, norms, intentions and hedonic responses in the selection of low salt bread in a longitudinal choice experiment. Appetite 7(2):127–139

Wood DF, Steward LM, Campbell CA (1988) Nutrient composition of 21 retail cuts of Canadian beef. J Can Diet Assoc 49(1):29–36

Wyant KW, Meiselman HL (1984) Sex and race differences in food preferences of military personnel. J Am Diet Assoc 84:169–175

Chapter 2

Cultural Determinants of Food Selection and Behavior

T. Johns and H.V. Kuhnlein

Cultural information and attitudes concerning food are communicated early in life and, because they relate to basic biological needs, can remain strong for those who fully participate in a particular culture. Such attitudes are known to persist throughout individual lifetimes and from one generation to the next. Conversely, dietary practices which are part of a culture are subject to continuous evolution. Most dietary problems of concern to nutritionists result from situations where cultural practices inhibit satisfaction of biological needs in relation to current environmental conditions. It is not surprising then, that change in cultural practices is central to the study of the relationships between culture and diet. This chapter is intended as a discussion of biocultural approaches to research on change in dietary practices which, it is hoped, will be useful to nutritionists interested in interdisciplinary work.

Most current research in nutritional anthropology examines social and culture variables jointly when considering dietary patterns (Pelto 1981; Axelson 1986). Sociodemographic and socioeconomic factors are mediated by culture and are important determinants of dietary behavior. However, this chapter concentrates on the concept of culture that is generally restricted to learned information, behavioral patterns and ideas.

Considerable applied research involving culture and diet focuses on cultural information as a predictor of dietary behavior, and as a possible tool for understanding how to initiate changes in behavior which will improve nutritional status. Since culture is learned, it can also be unlearned, or relearned, and education provides a route for intervening in dietary behavior. However, it is essential to understand the sociocultural milieu in which dietary practices developed before trying to change them.

Efforts to understand associations between culture and behavior do not necessarily offer insights into the origins or causes of particular habits. Data which correlate dietary behavior with a cultural characteristic (such as a food taboo based on religion) are unreliable as proof of true relationships without some ancillary form of verification. Full understanding of the way in which culture determines dietary behavior requires a more conscious attention to causal relationships and, invariably, a look backwards in time to understand the origins of cultural practices. Such studies may draw on historical or evolutionary data. Change, then, as an intrinsic element of evolution, is a fundamental factor in the causal relationships between culture and diet.

While applied research approaches to dietary change are generally concerned with contemporary areas of dietary maladaptation, more theoretical studies of dietary determinants can concern the adaptive nature of the interactions between culture and biology over generations or even millennia. An understanding of interrelations between culture and biology, both on theoretical and functional levels, is valuable for mediating dietary change. This chapter will consider theory and practice as interrelated approaches to the same problem.

Understanding Culture and Its Relationship to Diet

Both biology and culture affect what humans eat. Biological information, which is inherited genetically, and cultural information are different, although analogous in that they both contribute to how people meet their nutritional needs (Brown 1986). A key methodologic issue for scientists interested in cultural determinants of dietary behavior is to merge these two different types of information, as well as to merge the different methods and points of view of those disciplines that study culture and biology independently.

However, efforts to understand the effects of these forces on dietary behavior are complicated by difficulties in conceptualizing culture. The term "culture" is used in different ways and often is viewed holistically and in the abstract. Anthropologic debates concerning the concept of culture (see Sahlins 1976), while beyond the scope of this chapter, underlie the complexity of the issue. Much of the fundamental problem involves the interrelationship of culture and environment.

Somewhat concretely defined, culture is the pattern of knowledge, concepts, values, attitudes, beliefs and traditions that are learned and transmitted between individuals, often from generation to generation. However, it is difficult to reduce culture to basic elements or traits which can be measured directly (Brown 1986). Moreover, components of cultural patterns such as values and beliefs are complex, and not particularly concrete in themselves. It is difficult, therefore, to attribute human behavior to culture in a precise and measurable way, and we are forced to make approximations and delimit what we mean when talking about culture in a particular context. Adaptionist approaches emphasize the ways individuals and groups use culture to deal with the environment. Others consider cultural determinants independent of an adaptionist framework. Most contemporary research concerned with dietary behavior proceeds from the former point of view, which implies that biological need also influences culture.

Physical expressions of a culture can be easier to consider than cultural concepts and ideas. Eating habits or cuisine can be identified with a cultural or ethnic group, as can technologies used to procure food. While these manifestations of a culture convey information about that culture, they do not in themselves explain the variety of reasons for dietary behaviors.

It is difficult to study culture objectively. Rather we tend to interact with it and to experience it, as well as to observe it. Within our own culture we are participants and rarely observers; when viewing the cultures of others we invariably proceed from our own cultural framework. There is an inherent circularity in this, and even the careful application of the scientific method may not avoid the bias and value judgments that result from the uneasiness or uncertainty humans feel when they move from one culture to another. Each culture has its own logic, and to understand the logic of one cultural system by applying the logic of another cultural system may seem analogous to putting square pegs into round holes.

Traditionally, the discipline charged with the study of culture is anthropology. The standard anthropologic approach to studying the workings of another culture is for the observer to immerse him/herself in the life of that culture for a period lasting from several months to as long as several years (the so-called participant-observer method) after which s/he tries to explain that culture in its own terms (Wilson 1977; Bernard 1988). This is a useful approach for anthropologists, but it does not always provide objective information for the evaluation of dietary practices. Nonetheless, ethnographic fieldwork aims to provide a useful holistic, qualitative description of the dynamics of a particular situation (Jerome and Pelto 1981). Without the insights gained from such efforts, subsequent approaches can be lacking in common sense.

The participant-observer approach as practised by early anthropologists does not lend itself well to hypothesis testing. Nonetheless, as a means of understanding culturally determined dietary behavior, the participant-observer can provide useful insights that can then lead to more quantitative scientific methods. A key goal for those interested in refining ethnographic methodology is to combine participatory methods with more systematic efforts that can provide useful scientific information (Pelto 1981).

While traditional humanistic approaches to studying culture provide explanations of food choice, nutritionists usually need to put studies of the interactions of culture and dietary behavior on a more quantitative footing. Perhaps because scientific data regarding behavior and biology are more routinely quantified, those who study the interplay of culture with these concerns are obliged to quantify variables. In response, a semi-quantitative, but loosely defined methodology which parallels growth in scientific methods in anthropology has emerged in this field (Werner and Schoepfle 1987; Bernard 1988).

Scientific approaches that attempt to look at the dynamic relationship between culture and diet are challenged by the difficulties in precisely defining, in an operational fashion, the components of culture that affect specific behaviors. Since culture is really a sum total of many parts, it is not easy to gain insights into how the complexity of factors fit together to make up the whole. Part of this problem is that there are hundreds of cultures and great variation in characteristics within each culture. The apparent determinants of a particular dietary behavior may be relative to its cultural context, and the structure of the particular context may need to be established before variables capable of being studied are defined.

Faced with all the complexities of culture, it is not surprising that fully satisfactory quantifiable methods are uncommon. What has emerged in this field is an *ad hoc* methodology that quantifies what is quantifiable and generates hypotheses based on predictive relationships that can be tested using survey or experimental methods. Where quantification fails, traditional qualitative approaches continue to offer insights (Pelto et al. 1980). Nutritional anthropology, the umbrella for much of the research in this area, has been recently characterized in works by Jerome et al. (1980), Harrison and Ritenbaugh (1981), Johnston (1987), and Quant and Ritenbaugh (1987).

While anthropology and nutritional anthropology are viewed as the contemporary standard-bearers of research on the cultural determinants of food selection and behavior, other disciplines and subdisciplines have contributed substantially. These are sociology and medical sociology, nutritional epidemiology, ethnobiology, nutrition education, and community and international nutrition. Contributions have also come from food and nutrition specialists within the fields of geography, psychology, consumer economics, history, political science, food science, and other biological or social sciences. Each discipline has, for better or worse, brought its own methodology into the arena.

Some Qualitative Cultural Approaches

Traditional studies of food choice have primarily relied on historical and interpretive explanations. Grivetti and Pangborn (1973) categorized several of the approaches used in this field. Among those relying primarily on cultural explanations were cultural history and functionalism. Cultural history draws on data from archaeology, linguistics, history and oral history to trace the origins and spread of food habits. Functionalist approaches explain food choice in terms of the roles food plays in mediating group relationships (Foster and Anderson 1978). For example, food shared in meals or in the form of gifts is a important vehicle of communication. Type of cuisine is important in defining ethnicity and cultural identity, and within a group the types of foods consumed may reflect on social status. Other cultural determinants of food habits such as the symbolic significance of food and the importance of religion are widely recognized.

Cultural approaches and approaches that contain more biological and economic explanations to food choice are not exclusive; indeed several authors emphasize the need to integrate data from various sources (Fieldhouse 1986).

Biological Need, Adaptation and Behavior

The primary interest of nutritionists is that food meets biological needs. Culture may determine dietary behavior of individuals more than does biology; but, on the other hand, it may sometimes seem that people meet their nutritional needs in spite of culture (Sanjur 1982). Cultural patterns of food choice of human populations must be consistent with basic biological needs.

For those nutritionists interested in culture and food, cultural ecology provides a useful paradigm for explaining many of our dietary patterns. Cultural ecology is the study of the interactions of humans and environment in a cultural context. In this approach to studying diet and other aspects of human behavior, biology and culture are viewed as intercurrent forces in human affairs. Whereas nutritional anthropology includes the study of variables of nutrition (food intake and human health status) and culture, human cultural ecology in general includes the study of a greater variety of biological and environmental variables.

Cultural ecology has been criticized for providing only relative explanations for dietary behavior. It has difficulty in distinguishing between culture and biology as causal determinants of a particular behavior. For example, debate continues on the interpretation of the association of lactose tolerance with populations which engage in dairying (Armelagos 1987). Northern European cultures that engage in dairying have higher levels of the enzyme lactase than non-dairying populations such as the Chinese. Did cultural, social or ecological factors determine the difference in behavior and lead to evolutionary changes in the enzyme needed for milk digestion, or did an inherent difference in the levels of the enzyme allow one group and not the other to develop a dairying culture?

Studies in cultural ecology often lack a specific time-frame. However, explanations of the origins of many of the dynamic situations described can be made from a historical perspective. In a biological context evolutionary time is the ultimate scale of causality.

In an adaptionist paradigm, cultural behavior is viewed as playing a central role in the success of the human species. Cultural behavior enhances our ability to meet our biological needs in the face of changing environmental conditions. Biological evolution, which depends on genetic changes taking place over many generations, is a slow process, but culture is a revolutionary force allowing for more rapid adaptation to meet biological needs in response to changing environmental conditions (Brown 1986; Bryant et al. 1985).

Archaeological data suggest that the earliest cultural developments in the evolution of the human species arose as adaptations for food procurement. The earliest physical evidence of culture is in the form of stone tools left by hominids over one million years ago. Forms and wear patterns of stone tools support their role in processing animal and plant foods and in hunting (Keeley and Toth 1981; Toth 1985).

Developments of all cultural behavior, including that related to food procurement, were undoubtedly related to the development of intelligence. Current theory in the area of the evolution of the human diet suggests that food procurement was a driving force leading to increases in the size of the human brain (Milton 1988). In turn, the capacity for technological development has been a product of increased intelligence; it may also have been a feedback force leading to the physical changes necessary for the development of hand–eye coordination and in the structure of the human brain (Tobias 1981). Intelligence provides for improvements in processing and storing information, characteristics necessary for culture. The capacity to use language was a key development that furthered culture by allowing for the communication of information between cohorts and from generation to generation.

Technology and language were fundamental, culturally related determinants of food procurement that characterize the evolution of our species. Throughout history, techniques for manipulating physical objects have allowed us to procure and produce higher quality food. Language is a vehicle for communicating information of such techniques or other matters pertaining to food. In the modern world of food processing and the media advertising surrounding food, technology and language communication remain powerful determinants of dietary behavior.

While evolutionary history provides a useful focus for understanding the interactions of culture and biology in relation to diet, insights from this approach are unfortunately limited by the inadequacy of historical records. Studies of modern populations in conjunction with available data on history and environment enable one to test hypotheses and draw some conclusions as to how culture determines one dietary pattern as against another.

Models

Models give research on sociocultural determinants of food choice a conceptual and theoretical basis. Perhaps because scientific approaches to the cultural determinants of dietary behavior are attempting to merge abstract concepts into a rational framework, models which define concepts in a structured way are depended upon. Various understandings of food choice, such as those represented by the "multidimensional", "environmentalist", "ecological" and "motivational" models, have been described and recently summarized (Sanjur 1982; Fieldhouse 1986). Some are described conceptually, and others take the form of flow chart models (Jerome et al. 1980; Katz 1987).

Biocultural Models

The synthesis of concerns for biology and nutrition with approaches emphasizing culture has produced the so-called biocultural models of human food consumption. These generally have an ecological focus and assume on some level that food procurement systems should meet the essential nutritional requirements of a population.

Biocultural models consider a large number of factors that work synergistically rather than independently to influence dietary behavior. Individual consumption is determined by the biological needs of the individual, by environmental factors that determine human needs and the availability of resources, and by social and cultural influences. Biocultural models can generally be grouped into two types: some look strictly theoretically at causal determinants of dietary behavior; sociocultural models on the other hand, tend to be more applied in their ultimate purpose. Both types of model are concerned with the understanding of principles in such a way as to address specific nutritional problems of humans resulting from cultural and social forces. Theoretical models and sociocultural models each lead to particular ways of hypothesis testing. Each have specific goals, types of hypotheses proposed, and methods by which these hypotheses are tested.

The theoretical models look at the determinants of dietary behavior in very broad terms, often from the explicitly evolutionary perspective described above. They are characteristically proposed by biologists and biological anthropologists (e.g. Katz 1987). They take as a basic premise that food procurement systems must meet essential nutrient requirements so as to maintain and reproduce the population.

Hypotheses emerging from this approach tend to be tested by focusing on biological or archaeological data rather than sociocultural data. Variables are defined simply and concretely, perhaps in relation to cross-cultural surveys of biological or behavioral differences. Archaeologists place strong emphasis on technology, something for which a physical record of bones, pollen, tools and other artifacts often exists. In this approach there may be no need to quantify culture in any other way. Rather, relics of cultures are looked at comparatively between points in time or between geographical zones (Wing and Brown 1979).

The model of Jerome and colleagues (1980) is an example of a general sociocultural orientation to explaining dietary patterns. While incorporating aspects of individual needs and of environmental forces, the model places greater emphasis on social and cultural factors. Research in this field looks at these factors more precisely and more detailed models can be drawn.

Nutritionists and dietitians oriented in this way are concerned primarily with predicting the way sociocultural factors determine dietary behavior, rather than with elucidating generalized interactions between biology and culture. They would qualify the statement that food processing systems must meet essential nutrient requirements with a focus on the necessity to maintain the health of an individual or a population. These pragmatic approaches usually work in a contemporary time-frame, rather than in evolutionary time.

Nonetheless, change is a key focus in this approach as well. If culture in evolutionary terms is a mechanism for adapting to changing conditions, the pragmatic efforts of nutritionists are concerned with (although not necessarily explicitly), and involved in, the process of adaptation. Culture can be a force in resisting change, specifically through its function in non-biological aspects of food consumption, and in the short term cultural change (or lack of change) can contribute to poor nutrition and therefore be maladaptive. Rather than being an adaptive buffer mediating between biological needs and environmental conditions, when culture is out of step with change it can exacerbate the stress of change. Nutritionists need to understand the way populations and individuals respond to change and the social and cultural variables (such as knowledge acquisition) that determine it. Nutritionists also want to direct change in positive ways. They consider these issues at a social level. When nutritionists are concerned with a problem of dietary change in a contemporary group, it must be recognized that cultural history influences solutions that are possible for the present.

Biocultural models attempt to integrate insights gained from a multitude of sources into a unified and logical framework. They are a template for examining specific issues and cases of dietary behavior in more rigorous detail. They can be simply an explanatory device, but at their best they should generate testable hypotheses. Holistic models do not inhibit attention to detail; rather, they can allow for a reductionist emphasis in relation to the whole. Generally speaking elucidation of this detail, the generation of data, and the testing of hypotheses in relation to data, are the greatest needs for progress in this field.

Research Strategies

Hypotheses can be defined in general terms but in the research context must also be formulated in relation to specific problems. Operational definitions should be specified and variables defined that can be measured in a precise way. In order to

examine the association of cultural traits with dietary behavior in a predictive
way, cultural variables are used as independent variables in order to predict
variation in a dependent variable (in this case behavioral or biological/nutritional
variables) in relation to individual members of a cultural group. Once measurements relative to these variables have been made they can be followed by
appropriate laboratory and statistical analyses. Questions of the validity of
particular measures ultimately must be reexamined in relation to theory.

Pelto and colleagues (Pelto et al. 1980; Jerome and Pelto 1981) discuss
strategies for research in the area of nutritional anthropology which offer general
insights for the study of dietary behavior. They draw on methods of cultural
anthropology as well as on methods of nutritional epidemiology. Considerations
of the methodologic problems in establishing food intake can be found in these
references and in the 1981 volume of the Food and Nutrition Board of the
National Research Council.

Central to this approach is the focus of research on communities, although
exactly what one defines as the community depends very much on the goals of the
particular study. Samples should be taken in such a way as to represent the
behavior of the community over time (Jerome and Pelto 1981). Within communities, data are gathered about individuals; however, Pelto and co-workers (1980)
emphasize that attention must be given not just to mean data but to variability
among individuals. Although the dietary behavior of a community may follow a
specific pattern, culture determinants such as knowledge or beliefs may rest with
only a portion of a community, perhaps with women of a certain age (Bernard
1988). Individual behavior may be distinctly different from that of the community
as a whole. Additionally the transmission of culture is determined by dynamic
aspects of learning and behavior dependent on individuals (Cavalli-Sforza and
Feldman 1981).

As well as looking at specific cases, those interested in cultural determinants of
dietary behavior also want to identify generalized patterns of human behavior.
While valuable in this regard, cross-cultural studies may fail because of the
specificity of beliefs and concepts to a particular culture. Before generalities can
be made, concepts that are meaningful in different cultural contexts and which
show variation between cultures must be formulated. In addition, to distinguish
biological and cultural factors as determinants of diet, variables of culture and
dietary behavior should be examined in relation to groups which are faced with
the same environmental conditions and opportunities for food procurement. This
is difficult because it may not be possible to find two communities with the
necessary similarity in essential conditions.

Observations of human behavior are invariably gathered from "field" studies
among population groups. The possibilities do not exist for manipulating
variables in a controlled laboratory-type design. Quasi-experiments where some
intervention takes place may be possible in changing cultural practices. Situations
that lend themselves to evaluation as natural experiments, particularly where
conditions change over time, may occur opportunistically (Bernard 1988).

Verification is desirable where conclusions are based on correlation although,
in relation to cultural determinants of diet, it is not always possible. Ideally if
cultural behavior is a response to change, then verification of culture as a causal
determinant of diet takes place against the background of time. For community
nutritionists, change in culture that results in dietary change can verify the role of
a particular aspect of culture as a dietary determinant. Hypotheses can be tested

on the impact of educational intervention on change in dietary behavior (cultural change). This may take the form of an evaluation of particular efforts in nutrition education (Pelto 1981).

Secondary to the above approaches, experimental work on human bio-chemistry or nutrient composition of foods can provide supportive explanations for relationships between culture and diet. There are numerous examples in this area (Armelagos 1987; Katz 1987). The case of lactose intolerance and dairying introduced previously depends on information about the human enzyme, lactase, and on the carbohydrate composition of diets containing milk products. Geophagy is an example of a dietary behavior that appears to have both biological and cultural determinants. Experimental demonstration of the adsorp-tive capacity of edible clays eaten with bitter potatoes in North and South America supported the hypothesis that people engage in geophagy to detoxify the toxic glycoalkaloids in this food (Johns 1986). While this practice has a biological determinant, the choice and utilization of specific clays is culturally defined.

Such experimental approaches provide a way of verifying or negating con-clusions that are drawn from observation. They may provide evidence of causality or of origins although, as we see in the case of lactose, causality is not always simple to establish.

History and archaeology provide means for testing evolutionary theories. Evidence of cultural practices may be supported with evidence of food processing technology (Kraybill 1977). The effects of cultural practices on nutritional status can be apparent in human skeletal remains (Cohen and Armelagos 1984). Dietary change, particularly in the way societies adapt to changing environmental conditions and respond to technological change, is a key issue for anthropolog-ists. For example, using archaeological data, comparing strontium/calcium ratios in skeletal remains, Schoeninger (1982) demonstrated a shift to an increased reliance on plants among populations in the Middle East in the period preceding agriculture – 30 000 to 15 000 years ago.

Defining Cultural Variables

For nutritional anthropologists interested in studying the cultural determinants of food selection and behavior among human populations, the methodology used to define food use by individuals and populations has been more thoroughly described and characterized than has the methodology for defining quantifiable cultural variables (Food and Nutrition Board 1981; Center for Food Safety and Applied Nutrition 1987). Biological and behavioral variables might be defined relatively simply in relation to nutritional status and dietary intake but, as stated earlier, culture can be considerably more difficult to define operationally.

In defining cultural variables we want traits which are shared by individuals in a group. Culture, which by nature is shared by members of a group over time, cannot be understood by recording the experiences of a single individual, or even five or six individuals. While cultural phenomena can only be studied by gathering data from individuals, such data must be taken in a statistically sound framework.

The most straightforward cultural variables are those that emerge as commonly recognized differences. Manifestations of culture such as ethnicity (cultural or social group membership reflected in costume, cuisine, etc.), religion and language are easily agreed upon, well defined and relatively easy to determine. These inclusive concepts reduce the complex descriptive attributes of culture to a few discrete variables.

Ethnicity in particular, can be a predictor of dietary behavior (Axelson 1986; Rozin and Vollmecke 1986). Particular cuisines are associated with particular ethnic groups, and group membership means a strong likelihood that certain types of food will be eaten, perhaps in particular temporal patterns (Jerome and Pelto 1981). Religions often place constraints on diet and belonging to a particular religion is a likely predictor of some aspects of dietary habits. However, even these two obvious examples are sometimes defined with difficulty; for example, is an Indian defined as an Indian if one of four grandparents was not an Indian?

Cultural labels such as ethnicity may be intuitively predictive of dietary behavior and it is not always necessary to resort to quantification or statistics to appreciate their explanatory power. Nonparametric methods employing categorical variables, which can make use of these primary traits, are easy to work with. However, in spite of their fundamental simplicity and general predictiveness, these primary categories do not offer much depth for understanding the dynamics of the interaction between culture and diet. While most cultural variables are simple, specific instances of human behavior will be determined at greater levels of complexity. It is the complex details of culture which are difficult to measure.

Within Western societies, where definition of cultural variables is less problematic for scientists working within their own social milieu, considerable effort has been focused on knowledge and attitudes as predictors of dietary behavior (Axelson 1986). However, specific cultural concepts such as beliefs and values may not be comparable from culture to culture or between two subcultures. Nonetheless, in a particular context it may be a specific belief that is responsible for determining a dietary practice and study can focus on this. For example, the use of a fish oil rendered from ripened ooligans, *Thaleichthys pacificus* Richardson, is believed by the Nuxalk people of Bella Coola, British Columbia, and by other coastal Northwest Indian groups, to have healing powers for various disease states, including tuberculosis and constipation (Kuhnlein et al. 1982). "Ooligan grease" is incorporated into the regular family meal pattern as a condiment highly appreciated for its taste and healthfulness but, when required, it is also used in larger quantities as a medicine. Ooligan grease is labeled as an "Indian food" by the Nuxalk: you use it only if you are North American Indian, and if you are an Indian, you like it. One could test the hypothesis that native Indians use and like ooligan grease because they believe that it has medicinal value. Interviews could be carried out with a sample of coastal British Columbia residents with the following goals:

1.   To document that a significant portion of Indian people hold beliefs about the medicinal value of ooligan grease that are not shared by non-Indians.

2.   To collect dietary data that compares the use of, and preference for, ooligan grease between Indians and non-Indians.

3.   To establish a relationship between the use of ooligan grease and the holding of belief in its medicinal properties.

Dietary behavior is determined by multidimensional economic, environmental, biological and social variables in addition to culture. Because population data generated using epidemiologic models which correlate particular cultural traits with observed behavioral or biological variables do not establish a cause and effect relationship between these variables, investigators in this field are challenged to explain true causal relationships in such uncontrolled settings. In such situations sophisticated statistical treatments are a necessity. Also valuable is verification of results through ancillary sources of data such as those from experimental and historical studies.

Detailed collection of data on the prevalence and context of a range of cultural variables such as beliefs, attitudes or knowledge may assist in the understanding of a particular behavior. The complexity of these variables increases the need for statistical treatments. Multivariate methods of analysis may be useful under such circumstances. In theory, greater depth of understanding of dietary behavior may be the reward for the effort. For example, there may be considerable intra-population variation in the shared knowledge regarding a technology, and complex interviews and statistical methods may be able to look at the dynamics of the transmission of cultural information in depth (Cavalli-Sforza and Feldman 1981).

Other approaches to understanding dietary behavior can rely on recognition of aspects of culture that are less commonly treated in this regard. For example, consider the correlation of the structure and content of language with dietary behavior. Difference in taste taxonomies, for instance, have been associated with simple differences in behavior. The Malay language has detailed descriptors for salt in comparison with English (O'Mahoney and Muhiudeen 1977), with variations related to the source of the saltiness (e.g. "salty like sea water", "salty like salt" or "salty like soy sauce"). One assumes that salt has considerable importance in the Malay diet, even though data on salt consumption is not available.

Generally speaking, in making correlations between cultural variables and dietary behavior, more often than not "dietary behavior" is simply synonymous with consumption. However, other useful measures of diet-related behavior which may be culturally determined include those for food preference and acceptance, and taste preference, dependent variables which in turn may determine consumption. For example, cross-cultural (i.e. where ethnicity is the cultural variable) studies of taste such as that of Moskowitz and colleagues (1975) have correlated ethnic identity with variability in preference for basic sensory stimuli. Laborers in Karnataka region, India, had increased preference for sour and bitter in comparison with educated Indian students, whose taste preferences were consistent with those of most Western populations.

This study also gives an indication of the difficulty of attributing causality to culture–diet relationships. These authors hypothesized that dietary experience (frequency of previous consumption), not ethnic origin, determined taste preference. Even if a distinct ethnic difference could be supported, the effect of cultural versus genetic differences in determining ethnicity would not be readily distinguished. In this case it was probably social class rather than ethnicity that best separated the comparison groups.

Social patterns, such as community life, are general phenomena of biologically determined behavior but their specific expression in humans is culturally determined. In reality, culture may be equally, or more, determinant of social phenomena than is biology. In turn social pressures and perceptions can be an important vehicle for transmitting cultural values (Rozin and Vollmecke 1986). However, social status variables, such as income and education levels, often correlate more easily with strictly biological variables (such as hemoglobin or ferritin levels) which are more readily measured than the more difficult to quantify cultural variables (e.g. beliefs related to red meat consumption). Demographic variables that are used to define social structure (for example, aspects of family size and age structure in relation to "household composition" (Pelto et al. 1980)) do not reflect cultural determination in the same sense as do religion, knowledge or beliefs.

Political and economic factors are often broadly grouped with cultural issues since political and economic systems are manifestations of human culture. At the same time, political and economic conditions are affected by demographic, climatological and biological factors. Socioeconomic factors are powerful determinants of food choice, everywhere, but they may act essentially as environmental factors to which the individual adapts.

Widely used terms such as socioculture or sociocultural environment conceptually link social and culture factors, and at the same time make a distinction between them. Generally speaking, the methods of the traditional discipline of sociology are more quantitative than those of traditional anthropology. Recently measures of economic or social function have had considerable attention from nutritional anthropologists (Axelson 1986; Pelto 1981). While these determinants are often looked at "cross-culturally" they themselves are not cultural characters. Interest in attitudes, knowledge and similar determinants of dietary behavior has increased in the past 20 years (Axelson 1986).

If the goal is to understand the *cultural* determinants of dietary behavior the distinction between cultural and sociodemographic factors must be clearly made and the complexities of their interaction understood. The biocultural models discussed above attempt to embrace this complexity.

Multidisciplinary Approaches

Investigators interested in the field of the cultural determinants of dietary behavior emphasize the need for multidisciplinary research. Integrative models involve study of culture, behavior and nutrition. However, research cannot take place on such a general level, and to investigate any particular problem might involve the expertise of biologists (such as biochemists, nutritionists and ecologists), behaviorists and cultural anthropologists, and archaeologists.

Multidisciplinary research depends on more than just bringing together a team of persons from different fields of expertise. Principal investigators in this field must have a background that includes biological, behavioral and cultural sciences. What disciplines are called on to participate in a study may depend not simply on the nature of the research, but on the ability of the investigator to conceive of issues and ask questions that appreciate the feasibility of particular research and the availability of resources to carry it out.

Even when individuals have a degree of experience in several fields, it is hard to avoid a certain amount of naïvety when dealing with complex scientific concepts and methods.

Some effort must be made to integrate the different methods and perspectives of biologists and cultural anthropologists in the methodology of projects on food and nutrition practices. In this area an understandable uneasiness can exist between those who want to proceed according to strict scientific methods and principles and those who are uncomfortable with the restrictions and the more quantitative and technologic approach that this implies. A mutual suspicion can exist; on the one hand of technologic and reductionist approaches that attempt to measure and quantify for the simple sake of quantification, and on the other hand of approaches that are regarded as scientifically "soft". This struggle, which is sometimes manifest in the intellectual interchanges and personal relationships of biologists and cultural anthropologists, must be mediated successfully for progress to be made. Although anthropologists have been relatively late in adopting a hypothesis testing and data analysis methodology, rapid progress is being made in this regard (Werner and Schoepfle 1987; Bernard 1988) and is leading to productive research with biologists.

Conclusions

Current models provide an overall understanding of the relationships between culture and the biological need to procure food. Generally speaking, many patterns of culture are determined by biological forces, and in turn culture can be a determinant of dietary behavior. In order to understand the latter, we need to understand the former, and the adaptive evolutionary nature of the relationship between culture and biology over time.

The general patterns have been defined into models, but at this point greater attention needs to be focused on details of the interactions between cultural and dietary variables. Sociocultural studies provide details of specific interactions and will continue to be applied at this point of focus. Attention should be given to testing causal relationships through greater documentation of dietary change during periods of cultural change. Studies that look to the past will provide insight, but studies that correlate the two – modern and historical or archaeological – will be valuable.

Culture is difficult to define simply and difficult to measure. Thus we must continue to struggle with the concepts of culture and the definition of variables in such a way as to tease apart the complex dynamic nature of culture. The level of complexity at which cultural variables are defined will relate to specific research problems.

Statistical sophistication will be important in this task and in the understanding of cultural forces in the complex field situations where economics, politics, environment, biology, social factors and culture all act to determine dietary behavior and subsequent nutritional status. In this process of understanding, gradual refinement of rigorous scientific methods, of design and verification of hypotheses are essential.

At the same time we must be sensitive to the limits on our understanding and also allow for sufficient flexibility. Because of limitations caused by the complexity of culture, refinements of models will continue to rely on qualitative as well as quantitative approaches. Scientific understanding progresses both by means of knowledge gained through strictly defined methodologies and by seemingly more spontaneous insights. In a field where the subject matter is not easily defined and quantified, intuition will undoubtedly continue to be at the root of new insights. Where concepts are difficult to define, care must be taken not to overdefine them prematurely. Carefully laid out and disciplined methodologies generate new data most efficiently, but on the other hand, less structured approaches should not be discouraged. The most creative research in this field could incorporate both approaches.

With the apparent contradiction between need for greater rigor and greater flexibility a tension exists in this area of investigation. Hopefully it is a creative tension that will lead the discipline forward.

References

Armelagos G (1987) Biocultural aspects of food choice. In: Harris M, Ross EB (eds) Food and evolution. Temple University Press, Philadelphia, pp 579–594

Axelson ML (1986) The impact of culture on food-related behavior. Annu Rev Nutr 6:345–363

Bernard HR (1988) Research methods in cultural anthropology. Sage, Newbury Park, California

Brown PJ (1986) Cultural and genetic adaptations to malaria: problems of comparison. Hum Ecol 14:311–332

Bryant CA, Courtney A, Markesbery BA, DeWalt KM (1985) The cultural feast. West Publishing, St Paul

Cavalli-Sforza LL, Feldman MW (1981) Cultural transmission and evolution: a quantitative approach. Princeton University Press, Princeton

Center for Food Safety and Applied Nutrition (1987) Guidelines for use of dietary intake data. FDA, Department of Health and Human Services, Washington. FASEB, Bethesda

Cohen MN, Armelagos GJ (1984) Paleopathology at the origins of agriculture (editors' summation). In: Cohen MN, Armelagos GJ (eds) Paleopathology at the origins of agriculture. Academic Press, Orlando, pp 585–601

Fieldhouse P (1986) Food & nutrition: customs & culture. Croom Helm, London

Food and Nutrition Board (1981) Assessing changing food consumption patterns. National Research Council, National Academy Press, Washington, DC

Foster GM, Anderson BG (1978) Medical anthropology. Wiley, New York

Grivetti LE, Pangborn RM (1973) Food habit research: a review of approaches and methods. J Nutr Educ 5:204–207

Harrison GG, Ritenbaugh C (1981) Anthropology and nutrition: a perspective on two scientific subcultures. Fed Proc 40:2595–2600

Jerome NW, Pelto GH (1981) Integrating ethnographic research with nutrition studies. Fed Proc 40:2601–2605

Jerome NW, Pelto GH, Kandel RF (1980) An ecological approach to nutritional anthropology. In: Jerome NW, Kandel RF, Pelto GH (eds) Nutritional anthropology. Redgrave, Pleasantville, New York, pp 13–45

Johns T (1986) Detoxification function of geophagy and the domestication of the potato. J Chem Ecol 12:635–646

Johnston FE (ed) (1987) Nutritional anthropology. Alan R Liss, New York

Katz SH (1987) Food and biocultural evolution: a model for the investigation of modern nutritional problems. In: Johnston FE (ed) Nutritional anthropology. Alan R Liss, New York, pp 41–63

Keeley LH, Toth N (1981) Microwear polishes on early stone tools from Koobi-Fora, Kenya. Nature 293:464–465

Kraybill DR (1977) Pre-agricultural tools for the preparations of foods in the Old World. In: Reed CA (ed) Origins of agriculture. Mouton, The Hague, pp 485–521

Kuhnlein HV, Chan AC, Thompson JN, Nahai S (1982) Ooligan grease: a nutritious fat used by native people of coastal British Columbia. J Ethnobiol 2:154–161

Milton K (1988) Foraging behavior and the evolution of primate intelligence. In: Bryne R, Whiten A (eds) Machiavellian intelligence: social expertise and the evolution of intellect in monkeys, apes, and humans. Oxford University Press, Oxford

Moskowitz HR, Kumraiah V, Sharma KN, Jacobs HL, Sharma SD (1975) Cross-cultural differences in simple taste preferences. Science 190:1217–1218

O'Mahoney M, Muhiudeen M (1977) A preliminary study of alternative taste languages using qualitative description of sodium chloride solutions: Malay versus English. Br J Psychol 68:275–278

Pelto GH (1981) Anthropological contributions to nutrition education research. J Nutr Educ 13:S2–S8

Pelto GH, Jerome NW, Kandel RF (1980) Methodological issues in nutritional anthropology. In: Jerome NW, Kandel RF, Pelto GH (eds) Nutritional anthropology. Redgrave, Pleasantville, New York, pp 47–59

Quant S, Ritenbaugh C (1987) Training manual in nutritional anthropology. Am Anthrop Assoc, Chicago (Publication no. 20)

Rozin P, Vollmecke TA (1986) Food likes and dislikes. Annu Rev Nutr 6:433–465

Sahlins M (1976) Culture and practical reason. University of Chicago Press, Chicago

Sanjur D (1982) Social and cultural perspectives in nutrition. Prentice-Hall, Englewood Cliffs

Schoeninger MJ (1982) Diet and the evolution of modern human form in the Middle East. Am J Phys Anthropol 58:37–52

Tobias PV (1981) The emergence of man in Africa and beyond. Philos Trans R Soc Lond [Biol] 292:43–56

Toth N (1985) The Oldowan reassessed: a close look at early stone artifacts. J Archaeol Sci 12:101–120

Werner O, Schoepfle GM (1987) Systematic fieldwork (2 vols). Sage, Newbury Park, California

Wilson CS (177) Research methods in nutritional anthropology. In: Fitzgerald TK (ed) Nutrition and anthropology in action. van Gorcum, Assen, Amsterdam, pp 62–68

Wing ES, Brown A (1979) Paleonutrition. Academic Press, New York

Chapter 3

Biological Determinants of Food Preferences in Humans

H.J. Schmidt and G.K. Beauchamp

It's good food and not fine words that keeps me alive.

Molière, Les Femmes Savantes, 1672

As Molière obviously appreciated, we are biological organisms, and in order to survive we must eat. As omnivores, we must select a balanced diet from a myriad of possible foodstuffs that will satisfy our basic nutrient and energy requirements while avoiding ingestion of toxic substances. Metabolic and sensory mechanisms have evolved that contribute to our ability to manage this task, and some of our food preference patterns reflect these biological mechanisms.

An extensive literature using non-human models has identified a plethora of biological factors ranging from caloric and nutrient requirements, neurotransmitter and hormone levels, sensory hedonics, and genetic differences that appear to influence food choice patterns. The methods of investigating these factors in animals far outnumber those available for the study of the potential role of these same factors in human food preferences; hence the human literature is limited, somewhat fragmented and often contradictory. Whereas brain lesions, experimentally induced disease, illness and need states, and pharmacologic interventions are all common manipulations in animal research, they are obviously not all viable tools for the study of humans. In human research, the primary approaches to identifying biologically governed food preferences and choice behavior have been:

1. To investigate developmental responses to food related stimuli such as tastes and odors.

2. To explore individual differences under genetic control via twin studies.

3. To explore variations in food choice related to natural variations in health and nutritional status.

However, the mechanisms that underlie these phenomena are virtually unexplored. The possible role of hormones in human food preferences has been addressed in a limited fashion through investigations of variations of food choice over the menstrual cycle and in pregnant women. In addition, limited research has investigated the possible involvement of certain neurotransmitters in some human food choices.

This chapter will provide an overview of the aforementioned human research. In the developmental section, certain issues such as the effects of aging are ignored because they have been recently treated in detail elsewhere (Murphy et al. 1989) and the underlying mechanisms are so diverse that adequate discussion is not possible in the space available. Suggestions for future research will be offered when relevant and reference to animal models will be limited to those which have motivated human research or which suggest an avenue for future investigation.

Species-Typical Sensory Responses

Research concerning the ontogenic origins of species-typical (innate) reactions to food-related stimuli has primarily focused on taste. This is perhaps due to the central role of the taste system in regulating food acceptance–rejection and in initiating, through central nervous system (CNS) reflexes, digestive activity. In fact, it has been argued that the prototypical taste qualities (sweet, sour, bitter, salty and perhaps a few others) evolved to solve specific nutritional problems of obtaining sufficient calories and sodium while avoiding certain poisonous and corrosive plants (for discussion see Jacobs et al. 1978; McBurney and Gent 1979; Beauchamp and Cowart 1985).

Developmental studies of neonates and young infants provide strong evidence that the hedonic reactions associated with three of the four prototypical tastes – sweet, sour, and bitter – have an innate basis. The origin of hedonic response to salt, however, is less clear. Several experiments (Steiner 1973, 1979; Ganchrow et al. 1983; see Cowart 1981 for a review) have shown that infants with no prior feeding experience respond to sucrose solutions with vigorous sucking, lip licking, and expressions of contentment (however, see Rosenstein and Oster 1988), to sour solutions with lip pursing and nose wrinkling, and to bitter solutions with tongue protrusions, spitting and expressions of dislike. In contrast, salt solutions apparently do not elicit a distinctive gustofacial response in newborns although other studies suggest that they can detect salt (see Crook 1978). Sweet solutions are consumed by neonates in larger quantities than sour, bitter, salty or unsweetened solutions, lending further credence to the claim that sweet tastes are naturally preferred (Desor et al. 1973; Maller and Desor 1973). In addition, newborns tend to suck more avidly when presented with sucrose solutions than with water (e.g. Crook 1978). The consummatory and sucking patterns elicited by sour and bitter tasting substances are less consistent than for sweet, but by and large indicate an aversive reaction to some of these tastants, at least at high concentrations (see Beauchamp 1981; Cowart 1981).

If the newborn is initially indifferent to, or actually rejects (Crook 1978) salt, how then does salt come to be preferred in childhood and adulthood? This is an important question since salt is consumed by adults in quantities that far exceed biological need and has been implicated in causing or exacerbating hypertension (see discussion in Beauchamp 1987). A recent study of developmental changes in response to salt suggests that its appeal may, like sucrose, have an unlearned ontogenic component to its expression. Beauchamp and Cowart (1985) demonstrated a shift from indifference between water and a near isotonic salt solution to a preference for salt water (as measured by relative intake) at about four months of age. The consistency of the onset of this preference for salt, together with animal research which has documented post-natal increases in neural responses to salts (sodium chloride and lithium chloride) (e.g. Ferrell et al. 1981) has led Beauchamp and Cowart (1985) to hypothesize that there is a post-natal maturation of an unlearned preference. This is not to say that experience, perhaps in the form of lower or higher amounts of dietary salt, could not alter the preferred level of salt (see Harris and Booth 1987). Thus salt preference is likely to be characterized by an interaction between the inherent positive hedonic tone and individual experience with salty tastes (Beauchamp 1987).

Although odors are thought to be of major significance in determining food acceptability (Engen 1982), there is very little research on the origins of hedonic responses to odors, the mechanisms underlying these reactions, or the role odors may play in early food acceptance and preference. Several early studies of young childrens' reactions to odors failed to reveal adult-like hedonic reactions earlier than at five years of age (Kneip et al. 1931; Peto 1936; Stein et al. 1958; Engen 1982); this led to the claim that all hedonic reactions to odors are acquired through associational learning processes (Engen 1982). Recently, however, with the development of new methodology, adult-like hedonic reactions to a variety of food-related odors have been observed in children as young as three years of age, thereby challenging the empirical foundations of this claim (Schmidt and Beauchamp 1988). It remains possible that more sensitive measures will expose odor preferences and aversions in earliest infancy. Animal models suggest that mere exposure to an odor during a sensitive period in infancy can have a profound influence on preference behaviors expressed in adulthood (Woo and Leon 1987). Furthermore, these early odor experiences are related to stimulus-specific changes in the neural responsiveness of certain brain structures (Coopersmith and Leon 1984; 1986). The recent finding that human female neonates will selectively orient towards a novel odor (wild cherry or ginger) to which they have been exposed only briefly (Balogh and Porter 1986) provides impetus for additional human research in this area.

Although other sensory properties such as texture and appearance contribute to the perceived pleasantness of foods, there are no published data indicating whether such responses in humans are innate. Since the textural properties of foods can serve to help discrimination of edible from inedible substances, easily digestible from indigestible foods, and the macronutrient composition of foods, it is certainly plausible that textural food preferences, as taste, could have a genetic basis. Neonatal studies examining the relative acceptability of formulas varying in consistency would be one means by which to begin to explore this possibility.

Genetic Influences on Individual Differences

It has been seen that chemosensory responses to foods have an innate component. However, the fact that there is an unlearned relationship between a chemical attribute of a food and the response an organism makes to that chemical in no way implies that all individuals respond in exactly the same way. Individual differences are a hallmark of hedonic responses to foods as well as other sensory experiences. Such individual differences can be due to learning which is constrained by innately determined pathways. Salt taste preference apparently is determined by the interaction of innate mechanisms and individual experience as described above. Individual differences in choice and preference may also arise out of genetic differences between individuals.

Phenylthiocarbamide (PTC) provides an example of one such tastant: while the majority of the adult population (depending on ethnic group) experience PTC as extremely bitter, the remainder find it only moderately bitter or are completely insensitive to it. Comparisons of the taste thresholds for PTC in monozygotic and dizygotic twins reveal that sensitivity to this chemical has a significant heritable component. These differences in sensitivity to PTC and related bitter compounds have furthermore been associated with differences in the likelihood of acquiring food aversions; individuals who are more sensitive to PTC report more food aversions than individuals who are less sensitive to it (Fischer et al. 1961; Glanville and Kaplan 1965). Whether or not these sensitivity differences mediate aversions to specific types of bitter foods (e.g. coffee, grapefruit) has not yet been rigorously established (Kalmus 1971; Rozin and Vollmecke 1986), but the possibility has been acknowledged. Furthermore, it has also been suggested (e.g. Greene 1974) that an aversion to PTC-like chemicals serves to protect individuals against potentially anti-thyroid compounds. Differences in the proportions of individuals sensitive to this taste between cultural groups may be linked to differential selection pressures against consuming such compounds.

There is an olfactory analogue to PTC, namely the steroid androstenone. Approximately 50% of the adult population are not able to detect this odorant at very high concentrations, 15% find it somewhat pleasant, and 35% find it highly offensive (Wysocki and Beauchamp 1984). As with PTC, twin studies have demonstrated that these differences in sensitivity are, in part, genetically determined (Wysocki and Beauchamp 1984). Since androsterone can be found in significant quantities in pork products and some other foods, it is possible that future research will reveal certain food preference patterns to be mediated in part by hedonic responsiveness to this chemical. Consistent with this possibility, one study has reported a significant heritable component in preference for bacon (Krondl et al. 1983). More generally, there may be many odors for which genetically determined individual differences exist, and these differences too may contribute to human food choice variation.

Other genetic studies raise questions regarding the extent to which global food preference patterns will ultimately be accounted for by heritability factors. Some investigators have obtained a significantly higher correlation between global food preferences in monozygotic than in dizygotic twins (Krondl et al. 1983); however, Rozin and Millman (1987) have failed to confirm this effect with a sample of twins considerably larger than that of Krondl. While degree of preference for whole foods such as radishes, beef, liver and soft-boiled eggs could not be attributed to genetic predispositions in this latter study, the preferred degree of "hotness" from chili pepper did appear to have a heritable component. Greene and colleagues (1975) failed to find a significant heritability in preference for aqueous sucrose, lactose and sodium chloride solutions; this was in agreement with a small study by Beauchamp and Cowart (1985) using real foods. In addition, high intrafamilial correlations in specific food preferences have not been obtained (e.g. Birch 1980). However, the empirical studies on the genetic underpinnings of individual differences in food preferences have been limited in number and scope. Thus far, only a small sample of the range of possible foods has been investigated and the methods for evaluating preferences have not been exhaustive.

Need States, Metabolism, and Food Preferences and Aversions

While hedonic responses to some food stimuli have an innate underpinning, these are not always immutable: the nutritive state of the organism causes it to modulate the attractiveness of these foods. Indeed, it has been extensively documented that the pleasantness of various food tastes and smells can be affected by caloric state (Cabanac 1979). For example, the perceived pleasantness of sweet-tasting foods is substantially reduced following a high carbohydrate meal. Likewise, a greater positive hedonic response to proteins is observed when protein status is relatively lower than when following a high protein meal. These minor fluctuations in relative food preferences of normal individuals have been discussed in detail elsewhere (Cabanac 1979) and will not be elaborated on further. The subsequent discussion will focus on more radical alterations in food choices induced by deficiencies or excesses of certain nutrients due to dietary imbalances or metabolic abnormalities.

Patients with an inability to retain salt due to hormone deficiency (Addison's disease), may exhibit a heightened preference for salt, and consume it in large amounts. Wilkins and Richter (1940) reported a classic case of a pre-schooler who was hospitalized in order to control what his parents perceived as excessive consumption of salt. The child died shortly after being placed on a restricted salt diet in the hospital, and Addison's disease was diagnosed at autopsy. Presumably, the child's massive intake of salt was a form of life-sustaining self-medication. Heightened salt preferences have sometimes been noted in other patients with Addison's disease (Henkin et al. 1963), as has spontaneous consumption of large amounts of licorice (Cotterhill and Cunliffe 1973). Since natural licorice contains a sweetener, which in large amounts is capable of suppressing the symptoms of the disease, this preference has been interpreted as a nutrient-directed craving. The extent to which these salt and licorice preferences can be explained in terms of orosensory factors as opposed to metabolic feedback mechanisms remains to be explored. In fact, although there are many studies of sensory effects of sodium depletion in animal models, almost no such work has been done with human subjects (Beauchamp 1987). Species differences in mechanisms of salt appetite and expressions of salt preference make human studies necessary (Stricker and Verbalis 1988).

Another apparent case of nutrient-directed food choice is reported in a classic observational study of spontaneous food choices in newly weaned infants. Clara Davis (1939) reported that a child with severe rickets chose to drink small amounts of cod liver oil intermittently until the rickets had been resolved. These studies are of further interest because they suggested that infants could success- fully regulate diet when given an array of foods from which to choose. However, as has been noted previously (Rozin and Vollmecke 1986; Beauchamp and Maller 1987; Story and Brown 1987), the food choices were limited to "nutri- tious" items, with few foods containing high levels of salt or simple carbohydrates such as sucrose.

Several recent studies suggest that protein-calorie deficits are linked to specific protein (amino acid) flavor preferences (Beauchamp et al. 1987; Murphy 1987). Adults with biological indices of poor protein status prefer soup broth with higher concentrations of an amino acid additive, casein hydrolysate, than individuals with a higher protein status (Murphy 1987). Taste preferences in malnourished infants provide evidence that this effect may be independent of prior taste experience and not mediated by variations in cultural beliefs about the nutritive value of different tasting substances. Malnourished infants (2–24 months old) consumed more of a vegetable soup broth that had been laced with casein hydrolysate relative to the same soup without the additive. This preference pattern was reversed following recovery from malnutrition and in well-nourished infants without a history of malnutrition (Beauchamp et al. 1987). Since these consumption patterns were obtained in a series of 30-second trials that, in total, did not exceed 5 minutes, it is likely that they were mediated by chemosensory substrates as opposed to post-ingestional metabolic feedback mechanisms. However, the extent to which the association between the sensory properties of the amino acid–soup mixture and the nutrient deficiency is independent of experience, as it apparently is for the connection between caloric deficiency and sweetness preference, and whether it can be modified (e.g. Booth et al. 1982), remains to be determined.

Other than the aforementioned studies of salt and protein appetite, there is little further human evidence to link food preference patterns to specific need states. Animal models of diabetes, however, suggest another condition which may have a significant effect on food preferences. Rats with experimentally induced diabetes will select, when allowed, a diet that is high in protein and fat and low in carbohydrate. This pattern is in contrast to the diet-selection patterns of healthy rats (Tepper and Kanarek 1985; Tepper 1986). Since diabetic rats are not able to metabolize carbohydrates properly, this diet helps minimize the symptoms of the condition. The possibility that diabetes affects food choices in a similar manner in humans remains to be documented.

The body is not always wise. Lactose intolerance, an inability to metabolize the sugar lactose, is a condition the primary symptoms of which are illness, diarrhea, and gastric distress when milk products which are high in lactose are consumed. Despite the discomfort associated with its consumption, individuals apparently do not develop a flavor aversion to milk products although they may deliberately try to avoid products containing lactose (Pelchat and Rozin 1982). This phenome- non underscores an important distinction between preference behaviors and hedonics (Rozin 1979); in this case the avoidance of a food does not necessarily mean that the food is experienced as unpleasant. Similarly, nutrients that are unpleasant, such as medicines and vitamins, may be preferentially consumed, again demonstrating that food choice and food pleasantness are not necessarily connected.

Whether or not a particular metabolic abnormality has an impact on food preferences may be related to the dietary importance of the nutrient that is compromised by the disorder. Salt and protein are both essential for survival, whereas lactose is not an essential nutrient and its dietary absence does not seriously compromise the health status of an individual. Studies which investigate food preference patterns in the context of vitamin or mineral deficiencies (e.g. Greeley et al. 1986), in the context of food allergies, in an inability to metabolize certain amino acids, and in such diseases as celiac disease (Simoons 1981; Simmoons 1982), should provide further insight into metabolic control over patterns of food choice.

It is clear from the case of lactose intolerance that illness and gastric discomfort are not sufficient to produce an aversion to a food. The specific biological mechanisms that are involved in the acquisition of human food aversions remain elusive; however malaise appears to be an important factor (Pelchat and Rozin 1982). The study of conditions under which food aversions develop in patients who are undergoing chemotherapy for the treatment of cancer provides an excellent means for further exploration of this issue (see Bernstein and Webster 1980). Mattes and colleagues (1987a,b) have demonstrated that neither nausea nor vomiting are necessary to produce a food aversion in chemotherapy patients; the presence of one or both increases the likelihood of acquiring an aversion but does not guarantee it, and the administration of anti-emetic agents does not protect against the development of aversions. Since foods that are considered potent stimulators of gastric acid secretions (e.g. meats and fatty foods) are more prone to aversion formation than other foods (Mattes et al. 1987b) it is possible that the increased likelihood of esophageal reflux after ingestion of these foods may be an important physiological mechanism contributing to learned food aversions.

In summary, there is clearly an interaction between nutritional status and sensory response to specific foods. The nature of this interaction is complex however, and is dependent upon the particular combination of metabolic state and nutrient under consideration. In view of these complexities, the specific biochemical events involved in the control of food selection are difficult to elucidate as evidenced in the following section.

Neurotransmitters, Neuromodulators and Food Preferences

Hormones

Changes in taste preferences believed to be mediated by the ovarian hormones estrogen and progesterone have been extensively documented in non-human species (Wade 1976). Studies of taste responses in ovariectomized and pregnant rats, and rats injected with various hormones, indicate that high estrogen levels enhance sweet preferences and bitter aversions, while low estrogen levels (due, for example, to high progesterone levels as during pregnancy) reduce hedonic reactions to sweet and bitter substances. In addition, animals which exhibit more extreme hedonic reactions to tastes (e.g. when estrogen levels are high) select significantly less protein in their diets than animals which exhibit less extreme reactions (see Wade 1976 for a review). This observation has led to the hypothesis that in rats "hormone-related taste preferences have evolved as a hedonic mechanism to regulate protein intake during the varying conditions of nutritional requirement that accompany changes in reproductive status" (Wade 1976, p 260). The possibility that hormones exert similar effects on food preferences in humans is currently under investigation through studies of alterations in food preferences during the menstrual cycle and during pregnancy.

Despite many anecdotal reports that alimentary preference patterns fluctuate as a function of menstrual phase (such as cravings for chocolate and carbo-hydrates during the pre-menstrual phase), cyclic variations in dietary patterns or food preferences as a function of the menstrual cycle in humans have not been documented unequivocally. Although some studies have reported increased cravings for and consumption of carbohydrates, but not fats or proteins, during the luteal (pre-menstrual) phase when estrogen levels are relatively high, and not during the follicular (post-menstrual) phase (Dalvit-McPhillips 1983; Cohen et al. 1987), other studies, with more rigorous estimates of time of ovulation and more detailed dietary records, have failed to replicate these effects (Sophos et al. 1987; Tomelleri and Grunewald 1987). Depressed carbohydrate consumption follow-ing a glucose preload has been documented during the luteal phase but not during the follicular phase (Wright and Crow 1973; Pliner and Fleming 1983). Slower metabolism of carbohydrates during the luteal rather than the follicular phase has been offered as a possible mechanism for this phenomenon (Pliner and Fleming 1983).

In contrast to studies of the menstrual cycle, changes in food preferences have been clearly documented during pregnancy (Taggert 1961; Hook 1978; Walker et al. 1985; Finley et al. 1985). Approximately two-thirds of women report changes in taste and smell during pregnancy and concomitant changes in the desirability of various foods. For example, sweet, sour, and savory (salty) foods tend to increase in desirability, while coffee, tea, meats, and fatty foods tend to become aversive. Cravings for alien substances such as clay or ashes (pica) are also common. It has been suggested that these latter cravings are due to an increased physiologic need for certain minerals during pregnancy. While it is plausible to assume that these changes in food habits are caused by changes in physiologic requirements for specific nutrients, to date there is no research to confirm this.

Some research suggests that changes in food preference patterns during pregnancy may be mediated by changes in taste responsivity. Several studies have demonstrated decreased sensitivity to each of the four basic tastes during pregnancy (Schmidt 1925; Hansen and Langer 1935; both cited by Brown and Toma 1986). Brown and Toma (1986) have related decreased salt sensitivity in pregnant women to increased preferences for stronger salt solutions. Since animal studies have demonstrated that sodium deficiencies during pregnancy can be especially deleterious, it is possible that this change in salt sensitivity is related to an increased need for sodium during human pregnancy (Brown and Toma 1986). Although it is plausible that an increased requirement for energy during pregnancy might be mediated by a similar decrease in taste sensitivity and associated increased preference for sucrose, the data so far are inconclusive.

Brown and Toma (1986) report no change in preference or sensitivity for sucrose, while Dippel and Elias (1980) report a decreased preference for sugar during pregnancy. This latter pattern is in contrast to those reported previously, and is contrary to the prediction that an increased energy requirement would lead to increased preference for sugars.

The paucity of research on the relationship between human hormonal variations and food choices, and the absence of research on the mechanisms which govern such phenomena, render this a potentially productive field for further research. Prospective studies of pregnancy could provide particularly valuable data. In addition, direct measurement of hormone levels should prove more sensitive than the indirect methods of previous studies. Since fats are the richest source of calories, fat preferences during pregnancy should be investigated. It is interesting that studies of other species (rats), like some of the human studies, have also failed to observe qualitative changes in taste preferences as a function of the estrous cycle although intake levels may change (Wade 1976). This raises the possibility that normal hormonal fluctuations in non-pregnant women contribute only marginally if at all to variations in food choice.

Norepinephrine and Serotonin

Neurotransmitters, generally acting over shorter durations than hormones, have been implicated in controlling food intake and preference. Of particular interest in the last decade have been the neurotransmitters serotonin and norepinephrine and their effects on macronutrient selection. Non-human studies have shown that pharmacologic doses of norepinephrine (administered either centrally or peripherally) appear selectively to enhance carbohydrate consumption relative to protein or fat consumption (Leibowitz and Shor-Posner 1986). Agonists of norepinephrine (e.g. clonidine), administered either peripherally or injected directly into the paraventricular nucleus of the hypothalamus produce similar effects.

Serotonin (i.e. 5-hydroxytryptamine) in the medial hypothalamus appears to have the opposite effect, inhibiting carbohydrate consumption while sparing protein and fat intake. Central and peripheral injections of pharmacologic doses of serotonin or agonists of serotonin (such as D-fenfluramine) lead to a reduction of carbohydrate consumption while apparently not affecting protein or fat consumption. The opposite occurs when serotonin is reduced (see Blundell 1984, and Leibowitz and Shor-Posner 1986 for reviews). Studies of macronutrient selection following lesions of brain sites involved in the synthesis and reuptake of serotonin also suggest that serotonin plays a role in modulating carbohydrate consumption.

These animal data have motivated a limited amount of human research, exploring the possibility that the same neurotransmitters play a role in modulating human food preferences. Since it is impossible to conduct human studies involving central injections of neurotransmitters and other chemicals into specific brain sites, or the creation of specific brain lesions, only indirect measures are available to explore these mechanisms in humans. The two primary means for investigating the role of serotonin on food preferences in humans are to examine the effects of ingesting an amino acid precursor of serotonin – tryptophan – on macronutrient selection, and to study the effects of a serotonin agonist – D-fenfluramine (D-FF) (an anorectic drug used in the treatment of obesity) – on food consumption patterns. The rationale for these methods comes from rat studies which have shown that ingestion of tryptophan can increase brain serotonin levels and that D-FF has a selective effect on serotonin release and reuptake (see Blundell 1984 for a review).

Investigations of the role of norepinephrine on human food choice have focused primarily on the effects of noradrenergic drugs (e.g. clonidine) in the treatment of eating disorders such as anorexia nervosa. The results of such studies are difficult to interpret due to the complex nature of the disorder, and the majority focus on relative intake and not specific food preferences. Hence, the following discussion will be limited to investigations of the role of serotonin on food choice.

While several studies have reported that D-FF can reduce total food intake in humans, very few have investigated the effect of these substances on specific macronutrient preferences and the results to date are difficult to interpret (see Fernstrom 1987). For example, Blundell and Rogers (1980) demonstrated that humans treated with D-FF rated carbohydrate food items as less preferred than when not treated with the drug. However, their actual consumption patterns did not correlate well with these ratings; specifically, there was little reduction in intake when actual carbohydrates were offered to the subjects. One interpretation of this discrepancy is that the hedonic and sensory properties of foods prevail over the effects that D-FF can exert on food selections.

Other human studies have also yielded complex and contradictory results. Wurtman (1987) showed that when snacking is restricted to high-carbohydrate foods such as potato chips, fenfluramine-treated adults reduced carbohydrate consumption relative to adults treated with a placebo or with tryptophan. In another study, obese adults (carbohydrate-cravers) treated with fenfluramine also reduced their carbohydrate consumption during snacks relative to a baseline consumption rate without treatment. A third study, designed to investigate the relative effect of fenfluramine on protein and carbohydrate intake in obese patients, revealed a complex relationship between consumption of these two macronutrients and the nature of the meal. Although caloric intake was substantially reduced following drug treatment, snacks remained high in carbohydrates with little protein content. Main meals were affected in the manner predicted: protein intake was not statistically affected, while carbohydrate consumption was significantly reduced (Wurtman et al. 1981).

A major difficulty with some of these studies concerns identification of the macronutrients that might be affected by serotonin levels. Since the snack foods at issue tended to be high in both carbohydrates and fats, it is difficult to determine whether any intake decrement is best characterized as a reduction in carbohydrate, a reduction in fat, or both. In addition, since the snack foods contained almost no protein, floor effects may have obscured effects on protein (see Fernstrom 1987 for discussion). Furthermore, whether these effects extend to non-obese, non-carbohydrate cravers remains to be established.

Statistical shortcomings also mitigate the suggestion that carbohydrates are selectively affected by D-FF. In order to demonstrate a selective effect on carbohydrate consumption following fenfluramine administration, it is necessary to show a significant statistical interaction between the changes in carbohydrate and protein consumption; some studies, however, report only separate analyses of carbohydrate and protein. In these studies, protein consumption may in fact have decreased, as inspection of some of the reported data reveals, but whether there is a significant statistical difference between the reduction in carbohydrates and protein is not reported.

Human studies of the effects of tryptophan administration on food choice have also yielded conflicting results. Although tryptophan, in doses of 2 g or more, has been shown to reduce total caloric intake (e.g. Hrboticky et al. 1985), it is not clear whether this reflects a selective effect on carbohydrates. For example, admininstration of 2 g tryptophan was reported to reduce food intake by selectively affecting carbohydrate intake in relation to protein in one study (Silverstone and Smith, reported in Silverstone and Goodall 1984), but the same quantity of tryptophan failed to have a significant effect on carbohydrate consumption in a group of subjects in another study (Wurtman et al. 1981). The precise role of serotonin in macronutrient selection is further complicated by the observation that dietary depletion of tryptophan can cause a significant decline in protein selection, without affecting carbohydrate or fat selection or total caloric consumption (Young et al. 1988). Finally, whether normal physiologic variations in brain serotonin levels are sufficient to induce alterations in macronutrient selection remains to be determined (see Teff et al. 1989).

In sum, neither the specificity of the effects of D-FF or tryptophan nor the conditions under which such effects can be obtained in humans have been adequately articulated. It is interesting to note that the animal analogues to these human studies have also failed to reveal incontestable selective effects on carbohydrate consumption (e.g. Orthen-Gambill and Kanarek 1982; see Fernstrom 1987 for discussion). A major difficulty with many of these studies seems to be that the macronutrients under investigation, protein, carbohydrate and fat, each represent a diverse set of sensory experiences. Carbohydrates may be sweet or not, be salient in food or little noticed, and the same holds true for fats and proteins. Consequently, it may be unrealistic to expect humans' short-term regulation of these macronutrients to be under very rigorous control by specific neurotransmitter variation. In the long term, however, the postulated relationship may exist. Clearly further research is necessary to clarify this area of inquiry.

Endogenous Opiates

Other neurotransmitter-like substances, the endogenous opioids (morphine-like substances released in the body in response to a variety of environmental stimuli), are also important in regulating food intake and preference. Studies with rats now indicate that these substances serve as modulators of food and fluid intake. For example, small doses of opioid agonists may increase intake of palatable substances whereas antagonists have opposite effects. It has been suggested that effects on intake are mediated by alteration in the palatability of the food or beverage (Reid 1985).

Human studies have shown that opioid agonists increase food intake (Morley et al. 1985), whereas antagonists decrease intake during a single meal (Atkinson 1982; Thompson et al. 1983; Trenchard and Silverstone 1983; Cohen et al. 1985). However, evidence for a long-term suppression of food intake (i.e. weight loss) is apparently absent (Atkinson et al. 1985). In one study (Fantino et al. 1986), the antagonist naltrexone has been shown to depress pleasantness of glucose solutions and food odors, but not salt solutions.

The issue of whether changes in pleasantness or palatability of taste or flavors can account for the depression in food intake following opioid antagonists has received little attention. One recent study (Beauchamp et al. 1988) examined this question. Naltrexone treatment decreased pleasantness of a sweetened drink and salted soup and depressed food intake and hunger. However, the degree of food intake and hunger depression could not be statistically explained by the moderate changes in sensory pleasantness, suggesting that changes in food intake and flavor pleasantness may be separately influenced by opioid antagonists. They may not be causally related or have the same site of action. An interesting finding in this study was that the depression of intake following opioid antagonism was relatively less directed to carbohydrates and more directed to fats and protein. This is consistent with research showing that hunger reduction increases preference for carbohydrates relative to protein.

In summary, there is substantial evidence that opioid systems are involved to some extent in regulating human food intake and that these systems may modulate sensory pleasure associated with foods. However, the mechanism(s) of action and the interrelation of these variables remains to be elucidated.

Conclusion

A complete understanding of the biological factors underlying food choice and preference is a long way off. The work discussed above has shown at one level that this behavior is under genetic control and is influenced by nutritive state. At a more mechanistic level, variations in hormones and neurotransmitters have been linked to variation in food selection and preference. However, as yet, such variation has not provided particularly powerful explanations for the variety and richness of human food habits, choices and preferences.

Acknowledgement. Some of the work described here was supported by NIH Grant No.HL 31736 and the Fragrance Foundation Research Fund.

References

Atkinson RL (1982) Naloxone decreases food intake in obese humans. J Clin Endocrinol Metab 5:196–198
Atkinson RL, Berke LK, Drake CR, Bibbs ML, Williams FL, Kaiser DL (1985) Effects of long-term therapy with naltrexone on body weight in obesity. Clin Pharmacol Ther 38:419–422
Balogh RD, Porter RH (1986) Olfactory preferences resulting from mere exposure in human neonates. Infant Behav Dev 9:395–401
Beauchamp GK (1981) Ontogenesis of taste preferences. In Walcher DW, Kretchmer N (eds) Food nutrition and evaluation: food as an environmental factor in the genesis of human variability. Masson, New York pp 49–57
Beauchamp GK (1987) The human preference for excess salt. Am Sci 75:27–33
Beauchamp GK, Cowart BJ (1985) Congenital and experimental factors in the development of human flavor preferences. Appetite 6:357–372
Beauchamp GK, Maller O (1987) The development of flavor preferences in humans: a review. In: Kare MR, Maller O (eds) The chemical senses and nutrition. Academic Press, New York pp 291–310

Beauchamp GK, de Vaquera MV, Pearson PB (1987) Dietary status of human infants and their sensory responses to amino acid flavor. In: Kawamura Y, Kare MR (eds) Umami: a basic taste. Marcel Dekker, New York pp 125–138

Beauchamp GK, Bertino M, Engelman K (1988) Paper presented at the Tenth annual meeting of the Association for Chemoreception Sciences, April 1988, Sarasota, Florida

Bernstein IL, Webster MM (1980) Learned taste aversion in humans. Physiol Behav 25:363–366

Birch LL (1980) The relationship between children's food preferences and those of their parents. J Nutr Educ 12:14–18

Blundell JE (1984) Serotonin and appetite. Neuropharmacology 23:1151–1537

Blundell JE, Rogers PJ (1980) Effects of anorexic drugs on food intake, food selection and preference and hunger motivation and subjective experiences. Appetite 1:151–165

Booth DA, Mather P, Fuller J (1982) Starch content of ordinary foods associatively conditions human appetite and satiation, indexed by intake and eating pleasantness of starch-paired flavors. Appetite 3:163–184

Brown JE, Toma RB (1986) Taste changes during pregnancy. Chem J Clin Nutr 43:414–418

Cabanac M (1979) Sensory pleasure. Q Res Biol 54:1–29

Cohen II, Sherwin BB, Fleming AS (1987) Food cravings, mood and the menstrual cycle. Horm Behav 21:457–470

Cohen MR, Cohen RM, Pickar D, Murphy DL (1985) Naloxone reduces food intake in humans. Psychosomatic Med 47:132–138

Coopersmith R, Leon M (1984) Enhanced neural response to familiar olfactory cues. Science 225:849–851

Coopersmith R, Leon M (1986) Enhanced neural response by adult rats to odors experienced early in life. Brain Res 371:400–403

Cotterhill JA, Cunliffe WJ (1973) Self-medication with licorice in a patient with Addison's disease. Lancet i:294–295

Cowart BJ (1981) Development of taste perception in humans: sensitivity and preference throughout the life span. Psychol Bull 90:43–74

Crook CK (1978) Taste perception in the newborn infant. Infant Behav Dev 1:52–69

Dalvit-McPhillips SP (1983). The effect of the human menstrual cycle on nutrient intake. Physiol Behav 31:209–212

Davis CM (1939) Results of the self-selection of diets by young children. Can Med Assoc J 41:257–263

Desor JA, Maller O, Turner R (1973) Taste in acceptance of sugars by human infants. J Comp Physiol 84:496–501

Dippel RL, Elias JW (1980) Preferences for sweet in relationship to use of oral contraceptives and pregnancy. Horm Behav 14:1–6

Engen T (1982) The perception of odors. Academic Press, New York

Fantino M, Hosotte J, Apfelbaum M (1986) An opioid antagonist, naltrexone, reduces preferences for sucrose in humans. Am J Physiol 251:R91–96

Fernstrom JD (1987) Food-induced changes in brain serotonin synthesis: is there a relationship to appetite for specific macronutrients? Appetite 8:163–182

Ferrell MF, Mistretta CM, Bradley RM (1981) Development of chorda tympani taste responses in the rat. J Comp Neurol 198:37–44

Finley DA, Dewey KG, Lonnerdal B, Grivetti LE (1985) Food choices of vegetarians and nonvegetarians during pregnancy and lactation. Research 85:678–685

Fischer R, Griffin F, England S, Parn SM (1961) Taste thresholds and food dislikes. Nature 191:1328

Ganchrow JR, Steiner JE, Daher M (1983) Micronutrients and taste stimulus intake. In: Kare MR, Brand JG (eds) Interaction of the chemical senses with nutrition. Academic Press, Orlando pp 108–128

Glanville EV, Kaplan AR (1965) Food preference and sensitivity for bitter compounds. Nature 205:851–853

Greeley S, Stewart CN, Bertino M (1986) Micronutrients and taste stimulus intake. In: Kare MR, Brand JG (eds) Interaction of the chemical senses with nutrition. Academic Press, Orlando pp 108–128

Greene LS (1974) Physical growth and development, neurological maturation and behavioral functioning in two Ecuadorian Andean communities in which goiter is endemic. Am J Physiol Anthrop 41:139–152

Greene LS, Desor JA, Maller O (1975) Heredity and experience: their relative importance in the development of taste preference in man. J Comp Physiol 89:279–284

Harris G, Booth DA (1987) Infants preference for salt in food: its dependence upon recent dietary experience. J Reprod Infant Psychol 5:97–104

Henkin RI, Gill JR Jr, Bartter FC (1963) Studies on taste thresholds in normal man and patients with adrenal cortical insufficiency: the role of the adrenal cortical steroids and of serum sodium concentration. J Clin Invest 42:727–735

Hook EB (1978) Dietary cravings and aversions during pregnancy. Am J Clin Nutr 31:1355

Hrboticky N, Leiter LA, Anderson GH (1985) L-tryptophan on short-term food intake in lean men. Nutr Res 5:595–607

Jacobs WW, Beauchamp GK, Kare MR (1978) Progress in animal flavor research. In: Bullard RW (ed) Animal flavor chemistry. American Chemical Society, New York (Symposium series)

Kalmus H (1971) Genetics of taste. In: Berdler (ed) Handbook of sensory physiology, vol IV. Chemical Senses, London

Kneip HH, Morgan WL, Young PT (1931) Individual differences in affective reactions to odors. Am J Psychol 43:406–421

Krondl M, Coleman P, Wade J, Milner J (1983) A twin study examining genetic influence on food selection. Hum Nutr Appl Nutr 37:189–198

Leibowitz SF, Shor-Posner G (1986) Hypothalamic monoamine systems for control of food intake: analysis of meal patterns and macronutrient selection. In: Carruba MO, Blundell JE (eds) Pharmacology of eating disorders, Raven Press, New York

Maller O, Desor JA (1973) Effects of taste on ingestion by human newborns. In: Bosma JF (ed), Fourth symposium on oral sensation and perception. US Dept of Health, Education and Welfare, Bethesda (DHEW publication no. NIH 73–546)

Mattes RD, Arnold C, Boraas M (1987a) Management of learned food aversions in cancer patients receiving chemotherapy. Cancer Treat Rep 71:1071–1078

Mattes RD, Arnold C, Boraas M (1987b) Learned food aversions among cancer chemotherapy patients. Cancer 60:2576–2580

McBurney D, Gent JF (1979) On the nature of taste qualities. Psychol Bull 86:151–167

Morley JE, Parker S, Levine AS (1985) Effect of butorphanol tastant on food and water consumption in humans. Am J Clin Nutr 42:1175–1178

Murphy C (1987) Flavor preference for monosodium glutamate and casein hydrolysate in young and elderly persons. In: Kawamura Y, Kare MR (eds) Umami: a basic taste. Marcel Dekker, New York

Murphy C, Cain WS, Hegsted DM (eds) (1989) Nutrition and the chemical senses in aging: recent advances and current research needs. Annals of the New York Academy of Sciences, New York, vol 561

Orthen-Gambill N, Kanarek RB (1982) Differential effects of amphetamine and fenfluramine on dietary self-selection in rats. Pharmacol Biochem Behav 16:303–309

Pelchat JL, Rozin P (1982) The special role of nausea in the acquisition of food dislikes by humans. Appetite 3:341–351

Peto E (1936) Contribution to the development of smell feeding. Br J Med Psychol 15:314–320

Pliner P, Fleming AS (1983) Food intake, body weight and sweetness preferences over the menstrual cycle. Physiol Behav 30:663–666

Reid LD (1985) Endogenous opioid peptides and regulation of drinking and feeding. Am J Clin Nutr 42:1099–1132

Rolls BJ (1986) Sensory-specific satiety. Nutr Rev 4:93–101

Rosenstein D, Oster H (1988) Differential facial responses to four basic tastes in newborns. Child Dev 59:1555–1568

Rozin P (1979) Preference and affect in food selection. In: Kroeze JHA (ed), Preference Behavior and Chemoreception. Information Retrieval, London

Rozin P, Millman L (1987) Family environment, not heredity, accounts for family resemblances in food preferences and attitudes: a study. Appetite 8:125–134

Rozin P, Vollmecke TA (1986) Food intake and dislikes. Ann Rev Nutr 6:433–456

Schmidt HJ, Beauchamp GK (1988) Adult-like odor preferences and aversions in 3-year-old children. Child Dev 59:1136–1143

Silverstone T, Goodall E (1984) The clinical pharmacology of appetite suppressant drugs. Int J Obesity 8:23–33

Simoons FJ (1981) Celiac disease as a geographic problem. In: Walcher DN, Kretchmer N (eds) Food, nutrition and evolution: food as an environmental factor in the genesis of human variability. Masson, New York pp 179–199

Simmoons R (1982) Geography and genetics as factors in the psychobiology of human food selection. In: Barker LM (ed) The psychobiology of human food selection. AVI, Westport

Sophos CM, Worthington–Roberts B, Childs M (1987) Diet and body weight during the human menstrual cycle. Nutr Rep Int 36:201–211

Stein MP, Ottenberg P, Roulet M (1958) A study of the development of olfactory preferences. AMA Arch Neuro Psychiatr 80:264–266

Steiner JE (1973) The gustofacial response: observations in normal and anencephalic newborn infants. In: Bosma JF (ed) Fourth symposium on oral sensation and perception. US Dept. of Health, Education and Welfare, Bethesda (DHEW Publication No. 73–546)

Steiner JE (1979) Human facial expression in response to taste and smell stimulation. In: Reese HW, Lipsitt LP (eds) Advances in child development and behavior, Vol 13. Academic Press, New York pp 257–295

Story M, Brown JE (1987) Sounding board: do young children instinctively know what to eat? New Engl J Med 316:103–105

Stricker EM, Verbalis JG (1988) Hormones and behavior: the biology of thirst and sodium appetite. Am Sci 76:261–267

Taggart N (1961) Food habits in pregnancy. Proc Nutr Soc 20:35–40

Teff KL, Young SN, Marchand L, Botez MI (1989) Acute effect of protein or carbohydrate breakfast on human cerebrospinal fluid monoamine precursor and metabolite levels. J Neurochem 52:235–241

Tepper BJ (1986) Preliminary report: patterns of nutrient selection in long-term diabetic rats – association with diabetic complications. Nutr and Behav 3:181–193

Tepper BJ, Kanarek RB (1985) Dietary self-selection patterns of rats with mild diabetes. J Nutr 115:699–709

Thompson DA, Welle SL, Lilavivat U, Penicaud L, Campbell RG (1983) Opiate receptor blockade in man reduces 2-deoxy-D-glucose-induced food intake but not hunger, thirst and hypothermia. Life Sci 31:847–852

Tomelleri MS, Grunewald KK (1987) Menstrual cycle and food cravings in young college women. J Am Diet Assoc 87:311–315

Trenchard E, Silverstone T (1983) Naloxone reduces the food intake of normal human volunteers. Appetite 4:43–50

Wade GN (1976) Sex, hormones, regulating behaviors and body weight. In: Rosenblatt JS, Hende RA, Shaw E, Beer C (eds) Advances in the study of behavior 6:201–279

Wade J, Milner J, Krondl M (1981) Evidence for physiological regulation of food selection and nutritive intake in twins. Am J Clin Nutr 34:70–78

Walker ARP, Walker BF, Jones J, Verardi M, Walker C (1985) Nausea, vomiting and dietary cravings and aversions during pregnancy in South African women. Br J Obstet Gynaecol 92:484–489

Wilkins L, Richter CP (1940) A great craving for salt by a child with cortico-adrenal insufficiency. JAMA 114:866–868

Woo CC, Leon M (1987) Sensitive period for neural and behavioral response development to learned odors. Dev Brain Res 36:309–313

Wright P, Crow RA (1973) Menstrual cycle: effect on sweetness preferences in women. Horm Behav 4:387–391

Wurtman J (1987) Recent evidence from human studies linking central serotoninergic function with carbohydrate intake. Appetite 8:211–213

Wurtman J, Wurtman R, Growden JH, Henry P, Lipscomb A, Zeisel SH (1981) Carbohydrate craving in obese people: suppression by treatments affecting serotoninergic transmission. Int J Eat Dis 1:2–15

Wysocki CJ, Beauchamp GK (1984) The ability to smell androsterone is genetically determined. Proc Natl Acad Sci USA 81:4899–4902

Young SN, Tourjman SV, Teff KL, Pihl RO, Anderson GH (1988) The effect of lowering plasma tryptophan on food selection in normal males. Pharmacol Biochem Behav 31:149–152

Discussion

H.A. Guthrie

The motivation for studying the factors influencing food behavior is not merely to develop a conceptual, multifactorial model to describe its origin, but to understand the interrelationships between many influencing factors. This is critical in identifying appropriate points at which to intervene in order to influence the process and hence the outcome. The choices of these potential interventions, which can range from blocking taste receptors to changing attitudes about food, are determined largely by the perspective of the investigator. Most integrative models reflect the correlations between behavior and other measurable cultural and biological variables.

In the preceding chapters, by a social scientist, an anthropologist and a neurobiologist, it has become very evident that each of the disciplines represented has considerable interest in the study of food preferences. It is equally evident that each discipline views the subject from a quite different perspective and uses quite diverse methods in its attempts to elucidate the elements of food preferences.

Krondl describes the four major models that are used by the social scientist in studying the complexities of food preferences. While these involve a study of a great many social, sensory and environmental factors, these parameters have been kept to a minimum, and relatively simple, so that the resulting model is not as complex as many which attempt to delineate the multitude of forces which determine food behaviors. The essence of the models described is as follows: The *sensory model*, involving quantitative measures of taste preference under controlled conditions, recognizes that results are influenced largely by biological factors. The flexible *attitudes/belief model* suggests that behavior is best predicted by intention, which in turn is influenced by attitude. The *ecological model* used by anthropologists consists of six interrelated components of which biological requirements and psychologic needs are dominant and interact dynamically. The *food perception model* (the most complex) builds on sensory, attitudinal, environmental and demographic influences. In all of these models, the direction and weight of the relationships are important.

The chapter by Schmidt and Beauchamp introduces us to the intricacies of brain chemistry as it relates to preference for both salt and sugar. The authors demonstrate that the strength of a preference and the sensitivity to a flavor can be modified by a change in diet and, from that, the change in some of the neurotransmitters in the brain. They contend that food preferences that are genetically determined or influenced by nutritional state are more amenable to analysis and that food preferences can be modulated by hormone and neurotransmitter levels. Taste and possibly texture and smell as components of food acceptance are innately determined but can be modified by physiologic states such as nutrient deficiency and metabolic abnormalities, level of neurotransmitters and endogenous opiates.

Johns and Kuhnlein draw on their experience to provide illustrations of the strong cultural influences on food preferences. They discuss models of determinants of food choice which include biocultural approaches to the integration of concerns for biology and nutrition, with an ecological focus which is both theoretical and applied in orientation; a sociocultural model involving individual needs and environmental foci; and holistic models. Research methods from anthropology often provide insights into determinants of food behaviors. The authors emphasize the interaction among biological foci, patterns of culture and dietary behavior and the conflict between the need for both greater rigor and greater flexibility seen by investigators in the field.

All the contributors to this section recognize that they cannot explain food preferences solely from the perspective of their own discipline. The social scientists state that they need the insights of the physiologist and biologist in order to interpret many of the practices which are observed. Similarly the biology oriented contributors recognize that there is a strong social component in establishing and maintaining food preferences.

Throughout this section there are several recurring themes. Perhaps the most obvious theme is related to the complexity of the origins of food preferences, with contributors recognizing the many biological components relating to the need for particular nutrients, the avoidance of toxic substances in food, and the role which hormones and brain chemistry play in response to flavor and hence acceptance of foods. Many of the behavioral correlates of particular food practices reflect the social, cultural and environmental context in which food preferences are determined.

Among the many environmental factors which are recognized are characteristics of the population such as: educational level, age, economics, ethnic background, availability of food, access to sources of plants and animals, climate, water resources, proximity to stores, and availability of refrigeration. The food preferences of other family members have a definite influence either through their influence on attitudes toward particular foods or merely because attitudes and preferences of the gatekeeper of the family food supply influence what is made available to other family members. The biologists focus largely on the role of neurotransmitters and opioids in influencing the regulation of food intake both in terms of quantity and quality. It is, however, repeatedly emphasized that their discipline alone does not have the resources to fully explain or describe the origins of food preferences. The danger of naïve interpretations of data drawn from other disciplines is highlighted – a warning that should be kept in mind.

It is thus clear that a more complete understanding of the origins of food behavior will result only from a multidisciplinary approach. This requires that investigators in each discipline acquire an appreciation and understanding of the concepts, theories, techniques and vocabulary on which the others depend, and the limitations and strengths of various approaches. This section illustrates what is already a healthy intellectual exchange in which investigators who are looking at cultural factors as the independent variables are also looking for biological endpoints to explain their observations. Conversely others who measure biological factors as independent variables are seeking social, cultural and environmental determinants of the observed variance.

Another pervasive theme is related to the quality of social and biological data. There is reference to the "softness" of social science data and the "hardness" of biological data. A very objective look at the quality of the data is needed before we dichotomize it in this way. While it is reassuring to be able to put a sample of blood or urine into an analytical instrument to determine its biochemical components and to be able to replicate the measures within a relatively narrow range, it is simultaneously discouraging to learn that for many parameters the level of agreement from one laboratory to another in analyzing the sample is far from reassuring. The measurement of blood cholesterol, on which many people are now basing their dietary regime, is a case in point. Additionally it is learned that a dietary program should be based not on total cholesterol but on knowledge of cholesterol fractions. Frequently, when subjects are asked to make a rating on the "like–dislike" continuum of a Likert scale in order to quantify their feelings about food, misgivings about the replicability, or hardness, of the data arise. The chances are very good that it is equally as replicable as the majority of biological data, much of which could be easily modified by the subject merely adjusting his or her food intake before the sample was provided. Some biological parameters, however, are maintained within very narrow limits through an exquisite set of homeostatic mechanisms within the body. Before dismissing social data as relatively soft, it would seem important to take a very critical look at the quality and replicability of both types of data as the relative hardness of each may be about the same.

As one looks at many of the models used to explain the derivation of food preferences, one is struck by the amount of information that must be collected in order to fill in all the cells in the model before a meaningful multivariate analysis can be developed. Since much of the data must be provided by the subjects through test questionnaires or interviews, one cannot help but be concerned about the respondent burden and fatigue factor which could strongly influence the quality of the data to be analyzed. In contrast, in the case of biological data much of which is based on an analysis of blood, urine or other body tissues the additional information on other parameters can be obtained by additional analysis in the laboratory, without imposing any further burden on the respondent. Of course, the respondent does need to be informed of the analyses that will be undertaken and it is conceivable that any appreciable extension of the protocol will involve larger samples of tissue or tissue collected under specific conditions. In the USA investigators are often frustrated as the Office of Management and Budget monitors research protocols in order to protect study participants. This policy in the long run may do much to enhance the quality of the data.

The stability of both biological data and food preference and food intake data over time is a concern as findings are interpreted and attempts are made to reconcile divergent results from different investigators. There is sufficient evidence of the circadian variation in many biological measures to be fully aware of this as an alternative explanation for puzzling relationships. There is less documentation but considerable evidence that our reactions to many cultural and environmental stimuli are variable.

A disturbing feeling remains that there is a tendency to measure the variables that are easy to measure and then point to a myriad of other factors that could conceivably account for some of the variance as an alternative explanation. On the other hand, this is likely to be the best we can do until we have techniques sufficiently sensitive and feasible to measure these parameters.

The stage has been set for continuing productive interaction – the chapters in this section have recognized the need for learning the vocabularies of other disciplines and are aware of the application and limitations of techniques from divergent but similarly relevant fields of study.

With anything as complex as the determinants of food behavior, explanations may well be as variable as the pictures in a kaleidoscope – it depends on the perspective and the number of pieces involved. This analogy also points to the problem of measuring the impact of altering a single variable, as the interrelationships are so complex and interdependent that it is almost impossible to change only one factor without concurrently changing one or more other equally important factors. Care must be taken in identifying factors as major determinants without evidence that others which changed simultaneously were not equally responsible.

Section II
Measuring Behavioral Response

Introduction

N. A. Krasnegor

Scientists interested in establishing relationships between diet and behavior are faced with formidable methodologic challenges when posing and attempting to answer questions concerning such putative linkages. This part of the book is composed of three chapters and a discussion which were included to provide the reader with a perspective on the approaches used to identify, operationalize and measure behavior. The section has as its focus two main goals. The first of these is to elucidate ways of objectively defining and reliably and validly quantifying behavioral dependent variables in organisms that have consumed nutritional substances of interest. The second is to provide insights into the tactics and strategies that behavioral scientists employ in both laboratory and natural ecological settings to demonstrate and interpret significant changes in behavioral dependent variables.

Rodin's chapter, "Behavior: Its Definition and Measurement in Relation to Dietary Intake" reviews in general terms what is meant by behavior. The author provides an overview of the diverse dependent measures that are categorized under this general rubric and describes the principles used to choose among them and their parameters for the purposes of maximizing the capacity to discover diet–behavior relationships. Rodin illustrates her points by drawing upon examples from her own impressive research and that of other behavioral scientists on diet and behavior. She covers a wide range of issues that encompass the state-of-the-art of behavioral measurement, including strengths and limitations concerning conclusions to be drawn, problems relating to generalization, and differential strategies for laboratory as against field assessments. Her chapter articulates the complexities of correlational and causal modeling and the control procedures necessary to allow scientifically valid statements to be made about the effects of diet upon behavior.

The second chapter, "The Use of Animal Models to Study Effects of Nutrition on Behavior", by Crnic has as its focus the validity of animal models as an approach to measuring diet–behavior relationships. Crnic develops her work by describing how simple animal models must be modified with respect to both the species chosen and the variables included to demonstrate and/or discover presumed relationships. The author illustrates how findings from the animal research setting inform clinically relevant maneuvers for intervention in the care of patients. She similarly shows how findings from other research force changes in the complexity of models. Such alterations, the author points out, help to enrich the intricacy of questions that can be posed. This benefit however must be weighed against the costs associated with the difficulties in drawing straightforward conclusions about relationships between independent (nutrients) and dependent (behaviors) variables. Crnic demonstrates how as the complexity of experimental designs increases, there arises a corresponding requirement for an increase in the number and complexity of control procedures. Her work points out the value of animal models as heuristic devices for discovering both analogues and homologues in gaining an understanding of basic mechanisms that subsume nutrient–behavior relationships within both the model itself and the human condition that the model hopes to inform.

Thompson's chapter, "Behavioral Procedures for Assessing Effects of Biological Variables", complements the preceding chapter. The author provides a general review of the requirements for measures of human behavior within the laboratory setting. Thompson asserts that there are two main aspects of learning that should be measured in order to provide an evaluation of the potential and demonstrated effects of biological independent variables upon behavior. These are, respectively, short-term memory, and attention. The author employs examples from his own work, and that of others, in the domain of behavioral pharmacology to illustrate the reliability and validity of measuring behavior within the human performance laboratory setting. He reviews how procedures from animal laboratories (match-to-sample and repeated acquisition baselines) have been adapted to the human laboratory to measure the three aspects of learning deemed necessary for establishing valid relationships between biological and behavioral variables. Thompson makes a convincing argument for the procedures he describes having a generality not only between species but from the laboratory to relevant ecological settings in the natural environment. The baselines characterized in this chapter have a high potential for being adapted to research aimed at discovering diet–behavior relationships.

The section concludes with a discussion by Dews, who makes a number of salient points concerning mensurational issues. Dews points out that the most desirable characteristics of behavioral measures are that they ought to be both objective and quantifiable. He raises important cautionary notes about the use of statistical procedures for analyzing results obtained in behavioral studies. The author points out that diet and behavior studies can be conceptualized in a way similar to research conducted by pharmacologists. In keeping with this analogy, he argues that dose–response curves are required to gain a deep understanding of the fundamental relationships between dietary factors and behavior. Dews, as have others (see Crnic, Thompson), raises issue with the choice of traditional measures that have been used to assess behavior. In particular he is critical of the use of IQ to assess effects of malnutrition on behavior. Such measures are not usually helpful since they are relatively insensitive to changes in nutritional variables. The author also criticizes the term "learning" and suggests that a more neutral term which describes the performance should be substituted.

Chapter 4

Behavior: Its Definition and Measurement in Relation to Dietary Intake

J. Rodin

Introduction

One of the factors that makes the study of behavior both so difficult and so intriguing is that behavior is defined in many, quite different ways. None of these definitions is more or less correct than any other although some, as will be reviewed below, are easier than others to operationalize and yield more reliable data. Others may be more valid because they can be better extrapolated from laboratory observations to the "real world."

Some common definitions of behavior include: what people do, their attitudes and beliefs, their sensory experience and perceptions, and their judgments and feelings. The present chapter reviews these various definitions and how they are measured by both direct observation and by the use of verbal report. In nutrition research, behavior has been studied in both naturalistic and experimental settings by the use of a variety of research tactics including correlational and experimental designs. These various strategies will be reviewed in the present chapter.

One recent dichotomous categorization of types of behavior has helped to organize the various definitions and research strategies that have been used in behavioral research. In many ways, this distinction may serve to be especially useful in studies of diet and behavior. This distinction is one between *overt* and *covert* behavior. Overt behaviors are responses that are visible to observers (drinking and chewing are examples). Covert behaviors are unseen psychologic processes such as expecting, interpreting, or sensing, which are often postulated as intervening variables to make overt behavior more understandable (e.g. food preferences). As the study of behavior has matured, it has become clear that one often cannot understand actions without also understanding mental processes and feelings (for example, asking subjects to evaluate how pleasant various foods taste, or how hungry they feel often allows a better understanding of overt behaviors such as how much or how quickly they eat). On the other hand, sometimes the correlations between measures of actions such as amount eaten and inner states such as food preference is low. While this has sometimes puzzled and worried investigators, viewing them as two distinct types of behavior may help to resolve the apparent contradiction. If they are different, they may be influenced at times by different variables, both internal and external to the organism whose behaviors are being observed.

Both overt and covert behaviors can be viewed as a function of (i.e. caused by) two general classes of independent variables: one class includes the factors in an organism and the other comprises stimulus factors. Behavioral changes are thus dependent upon changes in one or both of these general classes of independent variables. Those who study human behavior refer to person, or dispositional, factors as the organismic factors. These include: personality traits, states of a person (such as hungry or satiated), and status characteristics (such as gender, age or race). Stimulus factors include variables such as aspects of the food itself, the setting and time of day, or social and cultural context. Often researchers will study the interaction of stimulus factors. Thus, depending on the context, the same amount of food can be perceived as appropriate or inappropriate; for example, a holiday meal as opposed to an ordinary weekday dinner.

Functional relationships can be expressed by the following formula:

$$B = f(O_t \times S_c)$$

It is read as: "Behavior is a function of organism factors at a given time in an organism's life interacting with stimulus factors within a particular context." When discussing properties of food, one needs to remember that eating food occurs in a particular environmental context, not in a vacuum, and the context affects the consummatory behavior dramatically. As the formula indicates, to understand any behavior, both organismic and situational factors should be considered.

Types of Independent Variables Most Typically Used

The study of people's eating behavior has sometimes focused on organismic factors (characteristics of the individual). These include: weight, gender, diet history, and stimulus factors. The greatest attention has been paid, however, to setting variables such as social factors, external cues and the characteristics of the food itself. Research has helped to make clear that people's nutrient intake is often dramatically influenced by factors extrinsic to an individual's specific nutritional needs at the moment, or to the nutritional value of the food itself. For example, the appearance, variety, and palatability of food stimulate intake as do the sight of other people eating and the type of occasion in which the eating behavior takes place (Striegel-Moore and Rodin 1986; Rodin et al 1989). The categories of independent variables that have been tested most widely are shown in Table 4.1.

Measurement Approaches

The variety and complexity of behavior poses major challenges to researchers who want to measure it. The first challenge is how to access the behavior(s) of

Table 4.1. Typical independent variables

Social and cultural factors
Setting/context
External cues
Variety
Cognitions
Palatability
Nutrient-related variables
Energy density variables
Texture
Volume
Liquid/solid
Characteristics of the individual

---

interest. Although some behaviors can be observed directly, many cannot. Thus, the task is often to make the unseen visible; to make private experience public. Many methods are available to do this, and each of them has its particular advantages and disadvantages. Two major methods of measuring behavior – observations and verbal reports – are considered below.

Observational Methods

Direct Observation

One major approach for assessing behavior is to employ real-time observational techniques. A direct observation is one that can be made with the naked eye. The behavior under investigation is clearly visible and overt, and can be easily recorded in writing or on videotape. For example, in a study of food choice, an experimenter could enumerate which foods subjects chose first; to assess preference he/she might record their facial expressions as they tasted the food. Steiner's format for rating the facial responses of newborns to different tastes is a prime example (Steiner 1979). This approach is especially useful for understanding hedonics or the experience of pleasure *vis-à-vis* food. A comparable, but more detailed system has been well worked out for rating the affective responses of adults (Ekman 1984); although it has not yet been applied in the diet and behavior literature it could make a fruitful contribution.

Naturalistic Observation. Observing some naturally occurring behavior with no attempt to change or interfere with it is called a naturalistic observation. Naturalistic observation is especially useful in the early stages of an investigation to discover the extent of some particular phenomenon or to get an idea of what the important variables and relationships might be. The data from naturalistic observation often provide clues to aid an investigator in formulating a specific hypothesis that can be tested by other research methods. For example, Stunkard and Kaplan (1977) observed diners at the same restaurant on nights when it served a buffet smorgasbord instead of dinner ordered from a menu. They noticed that more overweight people were in attendance on the smorgasbord night than during the rest of the week and that overweight clients ate more food on smorgasbord night, on average, than they did on regular menu nights. These data suggested that the amount of food available and the ease of consuming it may influence a variety of aspects of the behavior of overweight people, including whether or not they choose a particular restaurant.

Mediated Observation

As opposed to direct, naked eye observations, mediated observations require the use of special equipment or instrumentation. Such equipment allows more precise measurement of an observed behavior. In many studies one can observe behavior directly, using a variety of measurement techniques. These may include, for example, stopwatches, or computer-controlled timing devices to measure the behavior more precisely. Dependent variables may comprise such measures as: the length of each chew, the rate of chewing, the duration of time eating each food. In addition to allowing precise measurement, instrumentation can also help researchers to observe processes or events that are not visible to the naked eye, to ascertain neural and/or physiologic events. For example, by using the technology of the electrocardiogram (EKG), pleasure or preference might be assessed by determining a subject's physiologic arousal (assessed by monitoring correlated changes in brain electrical activity) that relates to various taste stimuli. Alternatively one might insert a balloon in the stomach of the subject and monitor abdominal contractions to assess hunger.

While some observations focus on the *process* of behavior, others focus on the *product* of behavior. The process of behavior is evaluated by, for example, measuring a subject's rate of eating, while the product of behavior can be determined by quantifying the differential amounts of food of each type eaten by a subject.

Verbal Reports

Researchers are interested in what people think, as well as in what they do. Alternatively, information might be sought about behaviors that are very difficult or impossible to observe (such as the intensity of binging and purging in bulimics). In these cases, investigators employ verbal reports. Subjects are asked a set of questions that require either written or oral answers. Sometimes these are taken at face value. Often though such answers are given differential weightings by researchers in light of nonverbal behaviors such as fidgeting or giggling during interviews. Verbal reports are also often the only method for getting information about subjects' beliefs, attitudes, and feelings.

In the nutrition field, the most typical type of verbal report data used is self-report of food intake. There has been an extensive dialogue in the literature regarding the weaknesses and advantages of records and recalls for periods ranging from 24 hours to 7 days. New methodologies for eliciting self-report reliably are being developed and will be highly useful to the field. In the author's own work, for example, a procedure has been validated in which subjects are instructed by telephone to complete a record for the 4 days preceding a face-to-face interview. Specifically, they are told to record everything that they eat and drink, trying to note the amounts as closely as possible. In some studies, when possible, subjects actually receive a training day in the laboratory to increase the reliability and validity of their record keeping. At the interview, these records are probed in detail. Portion sizes are established with the aid of measuring cups, glasses, bowls and food models.

Many have challenged the validity of verbal report data on the basis that they may not always be accurate or truthful. Subjects may give false or misleading answers for a variety of reasons. They may find reports of their true feelings or behavior to be embarrassing, they may want to look good and impress a researcher, they may not have observed accurately or remember clearly what they actually did in the past, or they may misunderstand the questions. Thus, researchers must make special efforts to ensure that the verbal reports they use are valid measures of the phenomena they are studying.

Questionnaires are the most common method of obtaining verbal reports. They can be framed in a variety of ways. These range in content from questions of fact (e.g. "How old are you?") to questions about past or present behavior ("How much do you diet?") to questions about attitudes and feelings ("How much do you like this food?"). The way in which questions are answered depends on how they are asked. Thus, fixed-alternative questions provide a limited set of alternative answers, and the subject picks the one that best represents his or her own position. The alternatives might be dichotomous ("yes or no"), or categorical (e.g. a 7-point scale from "strongly agree" to "strongly disagree"). Recently, a technique called magnitude estimation has been used. In this procedure, subjects assign a number to a particular alternative and then rate all others in relation to the first. Finally, open-ended questions may be used. These do not provide any alternatives; rather they ask subjects to answer freely in their own words. A typical example of an open-ended question is: "What do you like most about this food?" Questionnaires may be worded to be less obvious in their intent but the open-ended question tactic runs the risk that different subjects will interpret questions in different ways.

Research Designs

The conditions under which an investigator measures behavior are critical parts of any research study. A research design determines the types of relationships that can be studied and the sorts of conclusions that can be drawn. There are several major types of research designs, but the two most widely used in behavioral research are the controlled laboratory experiment and the correlational field study.

The Controlled Laboratory Experiment

A controlled experiment is a research method in which observations of specific behaviors are made under systematically varied conditions. An investigator may manipulate one or more stimulus variables and observe their effects on one or more behaviors. In the simplest case, the form or amount of one stimulus variable is changed systematically under carefully controlled, often admittedly quite restrictive, conditions, and a response variable is measured to determine whether relationships exist.

To clarify and reduce the probability that factors other than the independent variables are responsible for observed effects, experimenters attempt to control, or at least account for, other extraneous conditions. Three main ways are employed. These are: (a) the use of experimental and control groups; (b) the random assignment of subjects to experimental and control groups; and (c) the use of controlled procedures.

The latter method attempts to hold constant all variables and conditions other than those related to the hypothesis being tested. In diet and behavior research, instructions, temperature, time of day, how much food is available, and the context in which the study is carried out are all examples of control factors.

Examples of Controlled Laboratory Experiments

Most experiments in diet and behavior research include more than one independent variable because there are usually many factors that can influence the complex behavior(s) that underpin food intake.

Effects of Priming. Consider a study undertaken on the role of priming – providing a small taste of a good food and judging its effect on subsequent food intake (Cornell et al. in press). In this study, one of the independent variables was the type of priming food used – pizza vs ice cream (to control for palatability and fat content but to vary intensity of olfactory cues). A "no prime" control group was tested to determine the baseline amount eaten under the circumstances of that particular laboratory context, but without the presence of a priming stimulus. The other independent variable was an organismic factor – the subject's degree of hunger or satiety. To operationalize hunger or satiety, subjects in the satiety condition were fed a fixed amount of food after an overnight fast. Subjects in the hunger condition had the same overnight fast duration but were not fed. To try to minimize the role of other organismic factors on the manipulation of interest, the experimenters matched subjects on their degree of overweight, dieting histories and other weight-relevant variables.

The dependent variable in this study was the amount of pizza or ice cream eaten when a large amount of both were simultaneously offered after the initial prime. Another dependent variable was subjects' ratings of how much they desired to eat pizza or ice cream. The findings showed that priming with pizza or ice cream significantly increased the subjects' desire to eat the priming food. Further, when they were presented with both foods, subjects ate more of the priming food. While hungry subjects ate more overall than sated subjects, there were no interaction effects. In other words, being hungry or sated did not differentially influence the effects of priming on intake.

Sweetness and Food Intake. Recently Blundell and colleagues (1988) have taken researchers to task for inadequate designs that investigate the effects of various sweeteners on food intake. In certain cases sweetness in a food or drink is associated with high calories; in other circumstances sweetness may be present in the absence of calories. One important issue is to determine the conditions under

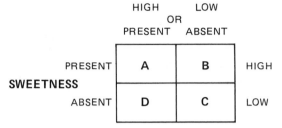

Fig. 4.1. Balanced design for investigating the effects of calories and sweetness using a preload procedure. (From Blundell et al. 1988.)

which ingestion of sweet substances affects appetite via the sweet taste itself, the ingested calories (or lack of them) or some combination of these elements.

Blundell et al. (1988) offer an important suggestion for what constitutes an appropriate experimental design for determining independently the effects of sweetness with and without calories. A fully balanced design is one in which two factors – calories and sweetness – are combined in a 2 × 2 matrix (see Fig. 4.1). This quartered table conceptualizes the design for four conditions in an idealized experimental approach. The design permits comparisons between two substances of the same caloric value but differing in sweetness, and of similar sweetness but differing in caloric content. Using the notation in the figure, a minimum design would comprise three of the cells (A, B and C). The effect(s) of caloric differences can thereby be compared for the same sweetness level (A and B), and different sweetness values can be compared for the same caloric level (B and C). The example cited may be called a partially balanced design but, according to Blundell et al. (1988), it is also the minimum required to permit meaningful comparisons. In practice condition C is normally the vehicle for A and B; calories and sweetness are added to make A and sweetness alone to make B. Condition D can be formed by adding calories to C or by suppressing the sweetness in A.

Clearly if only two conditions are used (for example A and B), then the effect of the vehicle (C) will remain unknown. If the effect of the vehicle is strong then it could overcome the action of the supplementary calories (A) or the additional sweetness (B). The most problematic situation would be if the two conditions (A and B) produce similar effects and the contribution of the vehicle is not known. For example, if equally sweet samples, differing in calories, produced similar reductions in hunger ratings and in calories consumed, a conclusion could not be made with confidence. If the design did not employ condition C, the effect of the vehicle itself would remain unknown. Even if the vehicle has no caloric or nutrient value, such as water, its volume may affect dependent variables relevant to food intake.

Within-Subjects vs. Between-Subjects Designs

Subjects in the experiments described above experienced only one level of each independent variable. They ate either high or low sweetened foods, not both. This type of study utilizes a between-subjects design. In this approach organismic variables – biological or dispositional factors that the person might bring to the experiment – are not a measured independent variable. If it is possible, a better way to conduct many experiments is to expose subjects to all levels of a particular independent variable. In this way many unmeasured organismic variables are controlled for because each subject serves as his or her own control. For example, the experiment on priming could be changed to allow all subjects to be tested twice, once when hungry and once satiated. The effect of the independent variable would then be measured within each subject, by comparing each subject's intake when hungry with his or her own intake when satiated. Such an independent variable is called a within-subjects variable.

Reliability and Validity

In controlled laboratory experiments the major methodologic concerns are reliability and external validity. Reliability refers to the precision of describing causes or mechanisms from experimental investigations, while external validity considers the extent to which outcomes can be generalized to other circumstances and situations.

The issue of external validity is often a concern for laboratory studies in the field of diet and behavior. Studies may need to be conducted under natural circumstances to establish whether effects observed in the laboratory generalize to field conditions. Blundell and colleagues (1988) note that for determination of the sensory evaluation of foods, their olfactory, gustatory and tactile dimensions are given extensive field tests to establish acceptability. Such testing is carried out because it cannot be assumed that laboratory findings, obtained under highly controlled and often artificial conditions, will necessarily provoke a positive response in the consumer. It can be argued that the effect of other aspects of food on appetite regulation mechanisms should be similarly examined in field settings once laboratory tests have suggested or confirmed specific hypotheses.

One issue currently debated in the scientific community is whether studies are necessary in situations where people know they are using low calorie products. One can postulate that people might use the known energy savings from low calorie products as their excuse for increasing intake of other foods. Undoubtedly, many new synthetic additives will continue to be developed and used in the food supply. These additives require study to determine their effects on appetite, satiety and metabolism. Such testing must of course extend to naturalistic situations. Ultimately most behavioral scientists use laboratory and field research together to advance knowledge. In practice this means that approaches which satisfy the demands of both internal and external validity will have to be found.

Correlational Studies

It is not always possible to carry out hypothesis testing experiments. Sometimes the phenomena are too broad to be reduced to specific variables that can be manipulated by an experimenter, as in the cumulative effect of excessive deprivation. Practical reasons may preclude the manipulation of independent variables. For example, in studies of the effects of being born to overweight or normal weight parents, weight of the parents is not under experimenter control. Finally, there are sometimes ethical reasons for not doing controlled experiments. The study of extreme stress on food intake is a good case in point. To study such phenomena, behavioral investigators utilize other types of research designs.

The primary alternative design is the correlational study. This approach assesses the degree of relationship between variables.

Examples of Correlational Studies

Suppose a researcher measures how much stress subjects are experiencing in their lives and how much food they are eating. This procedure was followed by Shatford and Evans (1986) in a study of bulimia. They wished to learn whether the occurrence of significant life stress increases the binging of bulimic women. The investigators attempted to ascertain whether a relationship exists between stress and food intake and, if so, the strength of the relationship. Causality could not be determined in such a relationship since correlation does not provide a basis for inferring a causal direction. The correlation could reflect any one of several cause-and-effect possibilities.

Assume that an investigator finds a correlation between stress and food intake, as Shatford and Evans did. They observed that as stress increased, binging increased. One possibility is that stress causes bulimics to eat more. An alternative is that bulimics experience stress when they binge. Another possibility is that a third variable is actually responsible for the relationship observed. Possibly people with a certain personality style are more likely both to experience stress and to be bulimic.

Problems with Correlational Data

Where there is no control over who gets exposed to an independent variable, and the variable is not systematically changed across subjects who are randomly assigned to experimental and control groups, the chicken-and-egg dilemma is an apt description of the problem. A way out of the predicament is to do the experiment, and then use appropriate statistical models (e.g. regression analysis or canonical correlation), which control for the effects of multiple dependent and independent variables (see Cohen and Cohen 1983, for a discussion of these statistical techniques). Other statistical procedures such as causal modeling, of which path analysis is an example (Heise 1975; Joreskog and Sorbom 1978), often allow investigators to make causal inferences even from correlational data, thus getting around some of the inference problems just described. For example, Shatford and Evans were able to suggest that stress preceded binging, rather than the reverse causal direction, by using path analysis. Correlational data may be subjected to causal analysis only when the investigator has an a priori model to test. Despite opportunities provided by these newer statistical procedures, however, there are still many problems with correlational data.

Correlations may also be spurious, biased or false, because the data are collected in a way that allows for a selection of cases in one part of the relationship to affect the data in the other part. Zimbardo (1988) gives the following dramatic example: Arizona and Florida have the highest incidence of respiratory illness and arthritis in the USA. The conclusion that the observation is due to pathogens in their atmospheres would be an erroneous one. Most likely this relationship occurs because their populations include great numbers of elderly people. So, although correlational studies have the advantage of being able to establish relationships between variables that cannot be manipulated in experiments, the interpretation of these relationships has to be done with caution and with the help of appropriate statistics.

Choice of Designs

All research designs have their place depending on the level of knowledge in an area, the type of populations an investigator has access to and the specific research question being addressed. In their selection of research design, investigators may be informed by a formulation of the "stages of research" model used by several institutes of the National Institutes of Health (Greenwald et al. 1986). In their analysis, research designs include: Phase I (hypothesis development), Phase II (methods development and testing), Phase III (controlled intervention trials), Phase IV (research in defined human populations), Phase V (demonstration and implementation studies).

Typical Dependent Variables

The most frequently used dependent variables in diet and behavior research are listed in Table 4.2. The measures tabulated are divided into two categories; observable measures (objective) and verbal reports (subjective). Many studies collect multiple measures because some variables are viewed as mediators and others as outcomes. Often one must ask how the measures relate to one another, as well as to the independent variables.

Table 4.2. Typical dependent variables

Observational methods
Amount consumed
Rate of eating
Microstructure of food intake
Frequency of ingestion
Motivation for food
Physiologic responses
Verbal reports
Food quality or intensity: taste, texture, smell
Hedonic ratings
Hunger/satiety

Examples of Measures for Assessing Dietary Manipulations

Measures of Food Choice, Amount Consumed and Physiologic Responses

Assume a study in which the effects of various nutrients are compared. The hypothesis is that fructose preloads compared with glucose preloads differentially affect subsequent food intake because of different effects on glucose levels and insulin release stimulated by the loads (Rodin et al. 1988). The loads were lemonade-type drinks, equal in calories to control for the effects of energy density and made equal in sweetness by adding 0.0001 g aspartame to the glucose load to control for sensory and hedonic effects. The investigators also wished to learn whether overweight and normal weight subjects respond differently to these manipulations.

A large and elaborate buffet was provided for the subjects and food intake was measured. The benefit of this procedure is that it is naturalistic and thereby increases external validity. Further this approach allows subjects a large choice, thereby insuring that internal validity is not compromised by the investigator having chosen a particular food that the subject does not like. Figure 4.2 shows the calories consumed as a function of type of nutrient previously eaten.

As Fig. 4.2 shows, subjects who first drank the fructose-sweetened solution subsequently ate significantly fewer calories than subjects who drank a glucose-sweetened drink. Indeed the suppression by fructose of subsequent intake appeared even greater for obese than for lean subjects. What is the mechanism on which the differential behavioral outcome is based? One hypothesis relates to the different amount of insulin released as a consequence of drinking the two preloads. Data on this point is shown in Fig. 4.3. The results suggest that caloric input (the amount eaten) was dependent upon the amount of insulin released.

Finally, the question of mechanism can begin to be addressed by correlating the magnitude of insulin released in response to the preload to the amount of food

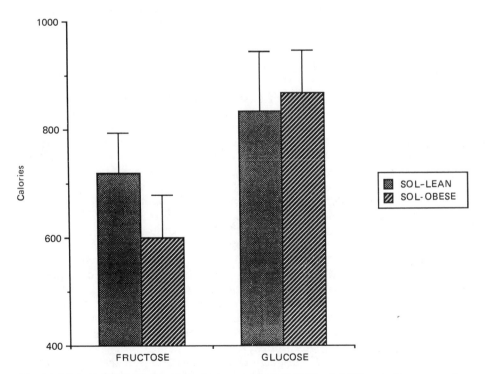

Fig. 4.2. Mean (± SEM) number of calories consumed at a buffet lunch 2.25 hours after consumption of 50 g fructose or glucose in solution. (From Rodin et al. 1988.)

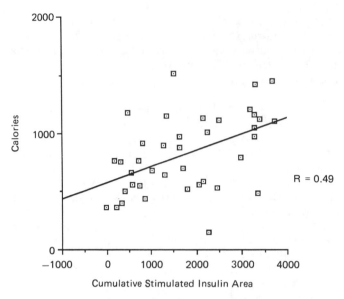

Fig. 4.3. Correlation between insulin responses to the consumption of 50 g of fructose or glucose and the number of calories eaten at a buffet lunch.

eaten. The data reveal that as insulin level increases the amount of calories consumed also increases.

Measures of Conditioned Hunger and Appetite

Another example of the benefit of multiple measures and the use of choice to measure food intake comes from a study of the role of arousal on the acquisition of a preference for food. Subjects were conditioned to the taste of a novel food (mango sherbet) after receiving either caffeine or placebo tablets (Rodin, in press). They showed strong differential effects on perceived hunger and food choice as a function of arousal, as is shown in Figs. 4.4 and 4.5. The data depicted in Fig. 4.4 show the relationship between learned hunger (conditioning to a novel taste) and appetite under conditions of arousal and non-arousal. Four groups of subjects were tested. The group labelled C-C received caffeine tablets prior to conditioning and the rating of their hunger on that session and the two that followed. Group C-P received caffeine prior to conditioning as well. But they received placebo prior to the performance session. Group P-C received placebo prior to conditioning and caffeine on the performance day. The group labelled P-P received placebo on both the pre-conditioning and performance days. The data reveal that arousal, induced by caffeine, during pre-conditioning produced an increased rating of hunger on the post-conditioning and performance sessions in groups C-C and C-P compared with the two control groups even though there were no differences in the groups prior to conditioning.

The data shown in Fig. 4.5 display the number of grams of sherbet consumed on the performance day by the four groups described above. Both of the groups who had been conditioned to the taste of mango sherbet under high arousal

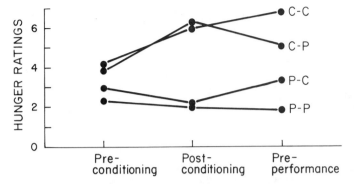

Fig. 4.4. Mean hunger ratings before conditioning, immediately after conditioning, and on a different day (performance) as a function of the subjects' arousal. C-C subjects had conditioning and performance trials after receiving caffeine; C-P subjects had caffeine for conditioning and placebo for performance; P-C subjects had placebo for conditioning and caffeine for performance; P-P had placebo on conditioning and performance trial days.

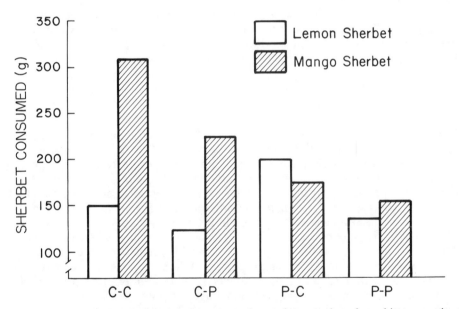

Fig. 4.5. Mean number of grams of sherbet consumed on performance day, when subjects were given lemon/lime and mango sherbet to eat. (Subject groups defined as in Fig. 4.4.)

during conditioning consumed more of that flavored food than the alternatively flavored dessert. The other two groups displayed no difference. Indeed they felt hungrier than they were at the pre-conditioning period although they had, as part of the conditioning procedure, just consumed several spoonfuls of sherbet. The effect of arousal on hunger was maintained on a second day (called pre-performance in Fig. 4.4).

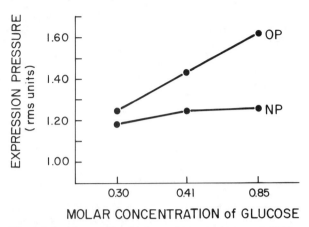

Fig. 4.6. Sucking avidity of infants of overweight parents (OP) or normal weight parents (NP) as a function of increasingly sweet glucose solutions. Amount consumed was held constant in all conditions. (Adapted from Milstein 1978.)

Choice, as measured by calories of each food consumed when both are present and freely available, is a strong measure of the effects of arousal on food intake. While presentation of a single food would have shown effects, the power of the inference is greatly increased by use of a choice paradigm.

The Microstructure of Eating

Milstein (1978) asked whether newborns would show different preferences for sweet taste as a function of the degree of obesity of their parents. Preference, indexed as avidity of sucking, was the measure of interest (shown in Fig. 4.6).

Avidity is defined as the rate at which the infant sucked. The experimenters controlled total intake so that all infants received the identical amount. This procedure controlled for post-ingestive differences as a function of differential sucking avidity. Although not displayed in Fig. 4.6, sucking for water was also tested, to provide a stronger inference regarding the effects of sweetness *per se*. The data in Fig. 4.6 show that infants of overweight parents are more responsive (suck more avidly) to the increasing sweetness of the solutions than the infants of normal weight parents. In this study infants did not differ as a function of their own fatness, body mass index, gestational age, or other developmental parameters. Note that other investigators have often inferred preference from amount consumed. Differences among individuals in amount consumed, however, may confound pure measures of preference. Thus sucking rate, with amount held constant across subjects, provides a stronger and clearer measure of preference alone.

Verbal Reports

Finally the use of verbal reports in diet–behavior research should be considered. Investigators have become quite interested in the relationship between post-

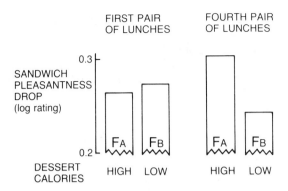

Fig. 4.7. Decreases in pleasantness of sandwiches (*ad libitum*). From the start to the finish of the course, before and after three pairings of flavors with starch augmented dessert (FA) and with low-calorie dessert (FB). Balancing of flavor–calorie pairings occurred across the group of eight participants. (From Booth 1985.)

ingestive signals and the regulation of food intake. Data from Booth and colleagues (Booth et al. 1976, 1982; Booth 1985) indicate that verbal reports of the pleasantness of food (sandwiches) taken as the first course, drop as a function of the caloric value of the dessert that follows after several pairings of the food (conditioning trials) with respectively high or low calorie desserts. Note that on the first trial, rated pleasantness of the food is independent of caloric value of the dessert that follows (see Fig. 4.7). After the fourth pairing of foods and desserts, the subjects rated the pleasantness of sandwiches followed by higher calorie desserts as much more pleasant than those followed by low calorie desserts. Notice the use of logarithmic transformations of the rating data, an often-used procedure in verbal report measures when the data are not normally distributed. This transform is especially of value when category scales are used. As an alternative to category scales, the method of magnitude estimation (see Bartoshuk and Marks 1986) is also often used. This method is especially powerful in within-subjects, repeated measures designs to circumvent problems associated with subjects having to remember the previously used number on the category scale.

Conclusions

By focusing on the rich array of behaviors described, and not just on amount of food consumed, and by taking account of organismic factors and of the aspects of the food in the context in which the food is eaten, investigators fulfill the most important methodological criteria for good diet and behavior research. There are many excellent studies in the literature but these have not been systematically reviewed since the task was to highlight methodologic procedures and issues. Thus in this chapter, the author cites her own research more often, because she is most obviously familiar with it. The emphasis is upon what has been most lacking in the field, especially the use of multiple measures of the same variable for convergent validity, and the use of field research to increase external validity of laboratory findings and naturalistic observations. The value of these approaches is that they can help to predict what real people will do under real-world circumstances. The richer, more varied and more precise the measures used, the closer the field will come to realizing this goal. Significant future advances can be made in the understanding of the relationship between diet and behavior by paying greater attention to methodological and statistical issues.

Acknowledgement. Preparation of this manuscript was supported by grant NS16993.

References

Bartoshuk LM, Marks LE (1986) Ratio scaling. In: Meiselman HL, Rivlin RS (eds) Clinical measurement of taste and smell. Macmillan, New York

Blundell JE, Rogers PJ, Hill AJ (1988) Uncoupling sweetness and calories: methodological aspects of laboratory studies on appetite control. Appetite 11:54–61

Booth DA (1985) Food-conditioned eating preferences and aversions with interoceptive elements: learned appetites and satieties. Ann NY Acad Sci 443:22–41

Booth DA, Lee M, McAleavey C (1976) Acquired sensory control of satiation in man. Br J Pyschol 67:137–147

Booth DA, Mather P, Fuller J (1982) Starch content of ordinary foods associatively conditions human appetite and satiation, indexed by intake and eating pleasantness of starch-paired flavours. Appetite 3:163–184

Cohen J, Cohen P (1983) Applied multiple regression/correlation analysis for the behavioral sciences. Erlbaum, Hillsdale

Cornell CE, Rodin J, Weingarten H Stimulus-induced eating when satiated. Physiol Behav (in press)

Ekman P (1984) Expression and the nature of emotion. In: Scherer KR, Ekman P (eds) Approaches of emotion. Erlbaum, Hillsdale

Greenwald P, Cohen J, McKenna J (1986) Phases of cancer prevention research development. Cancer Inst 79:389–400

Heise D (1975) Causal analysis. Wiley, New York

Joreskog K, Sorbom D (1978) *LISEREL IV: users guide.* International Educational Services, Chicago

Milstein RM (1978) Responsiveness in newborn infants of overweight and normal weight parents. Unpublished doctoral dissertation, Yale University

Rodin J Comparative effects of fructose, aspartame, glucose, and water preloads on calories and macronutrient intake. Am J Clin Nutr (in press)

Rodin J, Reed D, Jamner L (1988) Metabolic effects of fructose and glucose: implications for food intake. Am J Clin Nutr 47:683–689

Rodin J, Schank D, Striegel-Moore R (1989) Psychological features of obesity. Med Clin North Am 73:47–66

Shatford LA, Evans DR (1986) Bulimia as a manifestation of the stress process: a LISEREL causal modeling analysis. Int J Eating Dis 5:451–473

Steiner JE (1979) Human facial responses in response to taste and smell stimulation. In: Reese HW, Lipsitt LP (eds) Advances in child development and behavior, vol 14. Academic Press, New York, pp 257–295

Striegel-Moore R, Rodin J (1986) The influence of psychological variables in obesity. In: Brownell KD, Foreyt JP (eds) Physiology, psychology, and treatment of the eating disorders. Basic Books, New York

Stunkard AJ, Kaplan D (1977) Eating in public places: a review of reports of the direct observation of eating behavior. Int J Obes 1:89–101

Zimbardo PG (1988) Psychology and life. Scott, Foresman, Glenview, Illinois

Chapter 5

The Use of Animal Models to Study Effects of Nutrition on Behavior

L. S. Crnic

Introduction

The effects of nutrients on behavior, whether short-term or permanent, are of more concern in children than adults. This is because it is well established that the developing organism is more vulnerable to nutritional insult than the mature organism (Dobbing 1968). Indeed, permanent effects of nutrients on the brain and behavior may be possible only in immature organisms. Animal models of nutrient effects during development are more complex than adult models, and more difficult to extrapolate to humans. Therefore, the focus of this chapter will be on behavioral effects of nutrients during development, particularly permanent effects. However, most of the discussion will apply to the simpler case of the effects of nutrients on adult animals as well as to a variety of other perinatal interventions, such as exposure to drugs.

The Role of Animal Models in Science

Uses of Animals in Science

Animals fill many roles in the study of behavior. Denny and Ratner (1970) note that data on animal behavior may be gathered because we may be interested in the animal itself, in its economic value, in the relationship between animal behavior and evolution, and because one species may be a model for another. The function of non-human animals as a model for humans is the focus of this paper. Animal models can have several functions. First, the animal behavior process studied may be homologous to the human behavioral process of interest. In this case, application of the information gathered from the animal model may be straightforward. Second, the process studied in the animal may not be homologous to that in the human, but contain the rudiments of the human behavior, or simplified versions of the human behavior which make it useful in drawing general conclusions about the human behavior. Finally, the animal behavior may bear no resemblance to the human behavior, but may have been demonstrated as predictive of human response. The latter function of animal models is often employed in drug screening.

Advantages of Animal Models

Clearly, animal models of human problems are used because there are substantial advantages in doing so. First, animal models allow access to tissues not available in humans. Second, all aspects of the environment of animals can be controlled, and appropriate humane animal care minimizes variability in results. Third, dietary manipulations that are difficult or unethical to impose on humans can be imposed upon non-human animals. Fourth, in some cases, the animal model represents a simpler system which may be easier to understand than human behavior. Fifth, an animal with a much shorter lifespan than that of humans can be used. Thus, questions about development or aging can be answered in a reasonable period of time. Sixth, researchers may bring less bias to the interpretation of animal behavior than to the observation of human behavior. Clearly, some questions, such as those necessitating the sacrifice of the subject, cannot be answered without resort to animal models. These and other advantages of using animal models have been discussed at length elsewhere (Hinde 1976; Crnic et al. 1982; Smart 1984, 1987).

To make effective and valid use of animal models to gather data relevant to human behavior, one must understand (a) the scientific context in which animal research is embedded, and (b) the disadvantages and limitations of animal models.

Science as an Interactive System

To understand the role of animal modeling, it is useful to conceptualize science as an interactive system of approaches. Conceptualization of all approaches to science on any one dimension is of course impossible. However, to study the interactions among approaches, it is useful to devise a somewhat artificial continuum. One such continuum might be labeled "level of approach". This continuum ranges from the study of the smallest subatomic particles through chemistry and biochemistry to the study of cells, organs, organisms, the behaviour of organisms, how organisms interact, and proceeding on up to the level of complexity of sociology, economics, and international relations. Animal models can be seen as one point on this continuum.

Interactions Within the System

When various approaches to science are viewed as part of a system, it is clear that progress on any question in science is possible only when there is interaction between various parts of the system. For example, the study of a human behavioral problem must begin with observation of human behavior, and perhaps some experimentation with humans in order to define the behavioral question. Any further research is only as good as the definition of the behavioral question. If additional research requires invasive studies, control of the environment which is not possible with humans, or the imposition of procedures which are not ethically possible in humans, then it is appropriate to ask whether an animal model of the human phenomenon is possible. To ascertain this, knowledge from another part of the continuum must be used: basic research on animal behavior must be examined to determine whether there is an homologous system in animals that can be used to model the human phenomenon. Thus, animal modeling does not stand alone, but must be preceded by problem definition in humans and basic research on animal behavior. Similarly, the benefits of research from another part of the continuum, physics and chemistry, are tapped for instrumentation to study animal behavior. Further interaction within the system occurs when knowledge gathered from animal models is either applied to humans, or used to ask more sophisticated questions concerning humans. This interaction can be taken a step further: social scientists can be asked to employ their scientific knowledge to determine how best to change public policy or international relations to produce the desired effects on nutrition and behavior. Thus, although each scientist tends to specialize in research in one level of science, progress in both science and optimal care for humans requires interplay between many levels.

An example from the history of the study of the effects of malnutrition on behavior illustrates the value of interaction between various components of the system of science. It was clear early in the scientific study of malnutrition that it would be difficult to determine the effects of nutrient intake on humans. A major difficulty was that the nutrient intake of the subjects could not be controlled. Malnourishing human infants was unthinkable, but when naturally occurring malnutrition was studied there was variability in the nutritional conditions between individuals (reviewed in Grantham-McGregor 1987). Further, a variety of conditions which in themselves would be expected to be deleterious to behavioral development often accompany malnutrition (e.g. Chase and Crnic 1977; see Fig. 5.1). Thus, the effects of malnutrition acting alone could not be determined. Obviously, animal models were necessary and rodents (rats) were most often chosen for the model.

When behavioral psychologists studied malnutrition in rats they noted that its effects resembled those seen as a consequence of environmental isolation (Levitsky and Barnes 1972). This observation led them to ask whether environment interacted with the effects of malnutrition. The results revealed that environmental isolation did interact with malnutrition: isolation exaggerated the effects of malnutrition, while stimulation helped to ameliorate them (Levitsky and Barnes 1973; see Fig. 5.2). Scientists then reasoned that environmental enrichment could be used as part of the treatment for the effects of malnutrition in humans. This hypothesis was found to be valid, and interventions using enrichment of the environment were applied to humans with great success (e.g. McKay et al. 1978; Mora et al. 1981). McKay and colleagues (1978) provided, in addition to nutritional rehabilitation, variable numbers of 6-month periods of intellectual stimulation. As the number of these periods increased, IQ was shifted closer and closer to normal (Fig. 5.3). In addition, malnourished rat pups

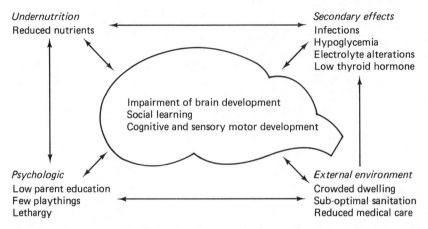

Fig. 5.1. Examples of the interrelations between the primary effects of undernutrition, its secondary effects, the influences of the external environment, and psychological factors on the brain and child development. From Chase and Crnic (1977).

(Massaro et al. 1974) and human children (Klein et al. 1975) were found to be less able than normals to make use of available environmental stimulation. These data led to the design of environmental interventions to actively engage the child.

The question then arose: What was the mechanism by which the environment had its effects? The animal model was again tapped to examine brain changes consequent to malnutrition and environmental enrichment (e.g. Im et al. 1976; Katz et al. 1982; Crnic 1983; Bhide and Bedi 1984).

Clearly, the interaction between research on malnourished humans and on animal models of human malnutrition has led to significant progress. Similarly, interaction of the study of human malnutrition with the fields of sociology and anthropology has been important. Those intervention programs that have made use of sociologic and anthropologic knowledge of various cultures have been most effective in intervening to improve nutrition and environment. In addition, understanding international relations plays an important role in famine control and thus prevention of malnutrition. This brief, and no doubt oversimplified, account illustrates how the interplay between basic research on animal behavior, animal models of human malnutrition and studies of malnourished humans led to progress in understanding the effects of malnutrition, and the creation of effective treatments for it.

Value of All Components of the Scientific System

Progress in the study of complex problems requires interaction among points on the continuum of approaches to science. Therefore, it is inappropriate to assign relative values to the various approaches to science. Rather, each should be viewed as providing a unique contribution to the whole process. Further, prediction of the eventual usefulness of any one field of research is not possible. Scientists need to do a better job of educating the public on this issue. When "Golden Fleece" awards are made for government funding of useless research, there is a tendency to pick on studies of the behavior of some obscure species that

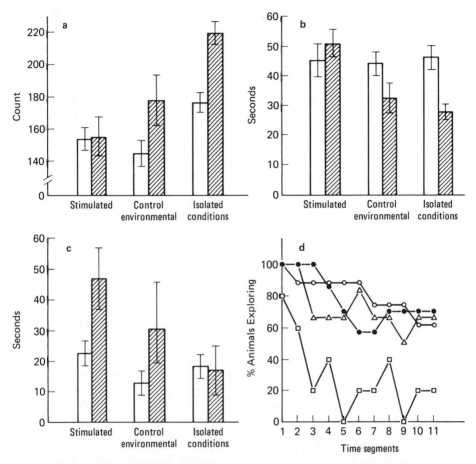

Fig. 5.2a–d. Mean and standard error of **a** locomotor, **b** following, and **c** fighting responses of rats on a low-protein diet (*hatched columns*) or a control diet (*open columns*). The exploratory response **d** is expressed as the percentage of animals within each group that entered the new environment per 5-minute time segment during the course of the test session. Well-nourished rats: ●, stimulated; △, isolated. Malnourished rats: ○, stimulated; □, isolated. From Levitsky and Barnes (1972).

has no obvious relevance or benefit to humans. Nevertheless, these studies may contribute to the solution of important human problems. A dramatic example which illustrates this point came from the laboratory of Howard Moltz. He and Michael Leon (Leon and Moltz 1972) discovered that infant rats can locate their mothers by scent. The signal that the rats detect is a derivative of deoxycholic acid present in the mothers' feces. Lee and Moltz (1984) discovered that the infant rats consume these feces, and that the pheromonal component, deoxycholic acid, served to protect the infants from enteritis. This finding led clinicians to the use of deoxycholic acid for the treatment necrotizing enterocolitis in human infants (Moltz 1984). There was no way of predicting that these studies of rat feces would have been so useful to humans. Thus, when all research is seen as part of a system, it is clear that one ought not to discount the value of any level of approach to science.

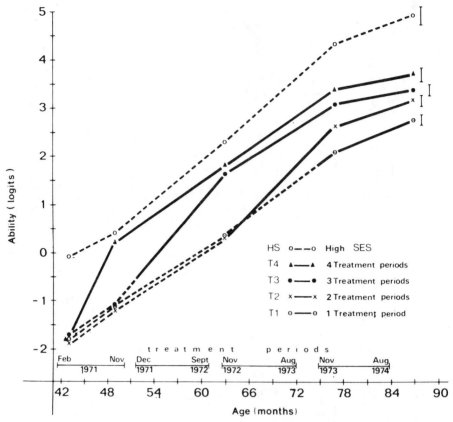

Fig. 5.3. Growth of general cognitive ability of children from age 43 months to 87 months, the age at the beginning of US primary school. Ability scores are scaled sums of test items correct among items common to proximate testing points. The *continuous lines* represent periods of participation in a treatment sequence, and error bars to the right of the curves indicate +1 standard error of the corresponding group's means at the fifth measurement point. At the fourth measurement point there are no overlapping standard errors; at earlier measurement points there is overlap only among obviously adjacent groups. Group T0 was tested at the fifth measurement point but is not represented in this figure, because its observed low level of performance could have been attributed to the fact that this was the first testing experience of the group T0 children since the neurological screening 4 years earlier. From McKay et al. (1978).

The Paradox of Animal Models

There would be no advantage in using animals to model human behavior if there were not differences between animals and humans which made greater experimental control possible. However, these same differences make risky the generalization of the results obtained with animals to humans. Thus, the value of a model lies both in its similarity to and difference from the system modeled. The goal of modeling is to strike a balance between the advantages of animal models and the price paid for the convenience of the model. Clearly, there are few perfect animal models for human behaviors: an animal complex enough to truly model human behavior could probably not be ethically used for study. As Hinde

(1976) notes: "When a model becomes an exact replica, it loses its raison d'être". As discussed in the next section, however, a model need not be perfect to be useful. In addition, it is impossible to be certain whether an animal model is a valid model for a human behavior. If enough is known to be able to determine whether the model is perfect, the behavior would have to be understood so well that further study would be unnecessary. Similarly, the suggestions that computers be used to model behavior cannot be taken seriously: if enough were known about the behavior to build a valid computer model, there would be no need for the model.

Limitations of Animal Models

The advantages of animal models do not come without a price. Those very differences between animals and humans that make animal models useful also make generalization between animal models and humans difficult. The seriousness of these differences depends upon the goals of the research. If one is interested in basic principles of development, differences between humans and animals may not be problematic. For example, even cultured cells can be used to ask such questions as whether neurons have the potential to change their neurotransmitter expression during development (e.g. Bunge et al. 1978). However, when the goal is to use an animal model to estimate the magnitude of an effect of a variable (such as malnutrition) the differences between the model and the system modeled can seriously limit the conclusion that can be drawn.

Relative Vulnerability

The most obvious and serious disadvantage of using animals to study the effects of nutrition is that they may not be as vulnerable to the behavioral effects of the nutrient as are humans. In humans, the more complex and sophisticated behaviors are those most vulnerable to insult. A non-human organism may not even be capable of such behaviors. For example, language is vulnerable to a variety of developmental insults (e.g. Lennenberg 1967), including malnutrition (e.g. Delicardie and Cravioto 1975): such effects could not be detected using rodents.

Non-human animals are not simply less vulnerable to nutritional insults than humans: in some ways they may be more vulnerable. First, one of the reasons that we use animals is that their life-spans are short compared with humans. Thus, their development is compressed into a briefer period of time. In rats, for example, all of fetal development occurs in 22 days. Nutritional intervention during even one of these 22 days represents a major portion of the fetal period, whereas a day would be an insignificant portion of human gestation. Second, most of the non-human animals used as models have multiple offspring. The rate at which the mother accumulates fetal tissue, calculated as the percentage of mother's weight gained per day of gestation, is 5 times greater in rats than in humans (Richardson 1973; Table 5.1). Thus, the rat might be expected to be more sensitive to nutritional intervention. This same comparison applies to the lactational period. Third, other animals are often born at a less mature stage of

Table 5.1. Accretion of weight per unit of mother's pre-pregnant weight per day of gestation

Species	Mother (wt. in kg)	Offspring (wt. in kg)	Gestation (in days)	Relative rate of accretion[a]
Mouse	0.04	0.014	22	7.6
Rat	0.20	0.045	22	4.9
Rabbit	1.90	0.400	32	3.1
Guinea pig	0.97	0.371	66	2.7
Macaca	6	0.45	164	2.2
Human	58	3.2	267	1.0
Pig	100	12.5	112	0.5
Elephant	3600	200	600	0.05

From Richardson (1973).
[a]Rates using human as standard = 1.

development than are humans. To the extent that the newborn is more exposed to environmental influences than the fetus, i.e. less buffered by maternal mechanisms, many species' offspring are more vulnerable at a less mature stage of development than is the human (Dobbing 1979). Fourth, the rate of development of these non-human animals is extraordinarily rapid in the vulnerable postnatal period (Fig. 5.4). Unfortunately, there is no one animal model which approximates the human on these variables. For example, mice and rats grow more

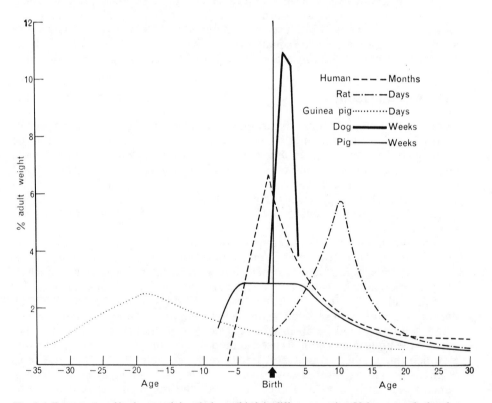

Fig. 5.4. Rate curves of brain growth in relation to birth in different species. Values are calculated at different time intervals for each species. From Dobbing (1968).

rapidly, are a larger burden to the mother, and are less mature at birth compared with humans. Non-human primates are a smaller burden on maternal tissues, but still twice that of humans, and are relatively more mature at birth.

Perhaps the most troublesome constraint on extrapolation from animal models to humans is the fact that there is no simple answer to the question of whether the model over/or underestimates effects of nutrient interventions early in life. This imprecision must be taken into account when extrapolating the results of work on nutrition. The most important point in this discussion is that while animal models can help elucidate general principles involved in nutrient effects upon behavior, they are not very good for answering questions about magnitude of effect. For example, animal models have provided important data about the types of intervention to use with malnourished children; however, the question of the extent to which malnutrition affects intelligence has never been answered using an animal model.

Simplicity

While studies employing animal models are clearly simpler than studies of humans, they are still extremely complex. An example from the history of the research on the effects of malnutrition on behavior illustrates some of this complexity, in addition to illustrating some of the issues involved in the choice of behavioral measures.

The most commonly used animals to study the effects of nutrients on behavior are rats. Early research attempted to study the direct effects of malnutrition on immature rat offspring (pups) by measuring learning deficits in the pups or in rehabilitated mature rats:

Learning was chosen as the outcome measure because it was thought to best approximate intelligence, a human trait held in high regard. Studies on the importance of early environment for later behavior revealed that this model was not adequate to measure the relationships of interest. Plaut (1970) was the first to recognize that one could not alter the nutrient status of the pups without having some effect upon an important aspect of the pups' early environment, the mother–infant interaction. This is because rodents are most vulnerable to nutritional insult early in development when they are dependent upon the dam for food and care. For example, the nutrient intake of pups can be altered by: (a) fostering large numbers of pups upon the dam; (b) altering the dams' nutrient intake; (c) removing pups from the dam for a portion of the day; (d) decreasing the dam's ability to provide milk by destroying some of her teats or removing some of her mammary tissue; and (e) removing the pups from the dams altogether and rearing them by hand. All of these methods have well-documented effects upon the early environment of the pups as well as their nutrition and have been discussed at length (e.g. Smart and Preece 1973; Massaro et al. 1974; Wiener et al. 1977; Crnic 1980).

There followed a period of time in which investigators attempted to devise a model in which alterations in the environment were minimized. This was an unobtainable goal because, to the extent that malnutrition alters the stimulus characteristics (sensory or behavioral) of the pups, it alters the behavior of the dam toward the pups, even if the dam experiences no intervention (Smart 1980). Thus the original model was modified:

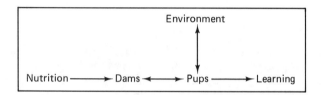

Further, evidence accumulated that a pup suffering from malnutrition was not as able as normal pups to make use of development-enhancing stimulation available in the environment (Massaro et al. 1974). This finding necessitated another addition to the model:

Environment

Nutrition ⟶ Dams ⟷ Pups ⟶ Learning

These new variables, while enriching the animal model, added to the difficulty in interpreting the results. However, this paradigm was an important step forward in gaining an understanding of the complexity of the effects of neonatal malnutrition in humans. The results led to the search for parallel findings in humans. For example, it was noted that nutritional supplementation of malnourished children had a major impact upon their interaction with their social environment. The supplemented children became more active and interactive and were perceived by their parents as more rewarding (Chavez and Martinez 1975).

A further complexity was realized when the outcome measure – learning – was carefully examined. Performance on a learning task is influenced by a variety of variables, including the motivation to perform the task, the emotional responsiveness of the animal to the testing situation, its motor ability to perform the task, capacity to pay attention to the task, and, no doubt, other factors in addition to the animal's learning ability.

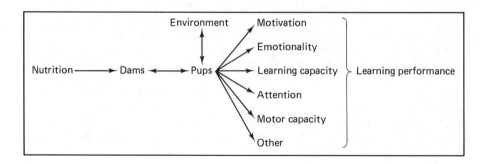

There are reasons to expect that early malnutrition might affect all of these intervening variables (reviewed in Levitsky and Barnes 1973; Crnic 1976; Levine and Wiener 1976; Smart 1977). For example, a rat malnourished early in life will have a permanently altered body composition and response to food deprivation and food reward (reviewed in Crnic 1984). The use of aversive stimulation to motivate behavior provides no solution: previously undernourished rats are more responsive to electric shock (Smart et al. 1975; Lynch 1976).

The complexity of assessing learning in previously malnourished rats, while troublesome, has led to important methodologic and conceptual advances. Researchers have been forced to increase the sophistication of their experimental designs to control for performance variables when studying learning. Control conditions were added to determine whether nutritionally treated animals differ from normals in their ability to perform the motor components of the task, in their sensory capabilities, their attention, and their motivation. These procedures are essential if conclusions are to be drawn about learning capacity. For example, if shock is to be used to motivate performance, a separate group of animals must be tested to determine whether their response to shock is the same as controls. One elegant approach is to use a task which lends itself readily to control conditions, such as the Morris maze. In this task rats must locate a platform hidden under the water in order to escape from the water (Morris 1984). The subject does so by using visuospatial cues in the room. Rats can first be tested for their ability to locate and swim to a *visible* platform. Performance when the platform is visible can be used to determine whether motor and sensory facilities are intact and whether the motivation to swim is normal.

The discovery of the complexity of animal models forced a reassessment of the choice of outcome measures for both animal and human studies. Clearly, the effects of nutrition on intervening variables such as motor performance, motivation and emotion are of interest in their own right. Although intelligence and the ability to learn are extremely important, they have been studied to the neglect of other important aspects of behavior. This occurred perhaps because intelligence is highly valued by those who conduct research. It is essential to study other aspects of behavior and motivation which can have an important impact upon the quality of life. For example, there is evidence that the poor performance of malnourished children on IQ tests appears to be due to their inattention and unwillingness to try to perform well (Cravioto and Delicardie 1979).

Some of the most successful work on human behavioral effects of malnutrition has employed dependent variables other than IQ performance. This work provides an additional example of the importance of the interplay between animal and human research. Barrett and Radke-Yarrow (1985) and Barrett (1987) used the results from animal studies to guide research which sought to determine which behavioral systems are affected by early malnutrition in humans. The results of animal studies were used to develop analogous behavioral measures for humans. They hypothesized that chronic undernutrition in school-aged children would lead to low exploration, withdrawal from competitive situations, poor frustration tolerance, poor impulse control, erratic activity level, low social involvement, timid or anxious behavior and infrequent happy affect. They used both naturalistic observation and structured laboratory situations to test these hypotheses. This resulted in a research program with a level of sophistication that far exceeds that of typical studies done in this field.

The work of Barrett (1987) cited above indicates that measures of behavior other than motivated learning might be more sensitive in detecting effects of insults (e.g. Strupp and Levitsky 1983). This may be the case because the ability to learn is so essential to an organism's survival that one would expect that such capacities would be relatively impervious to environmental insult.

Sensitivity

Animal models can be used to answer questions about the effects of malnutrition in humans that are separate from the effects of factors such as inferior housing or infectious disease. But in simple experiments, interactions, such as those seen between nutrition and environment discussed above, may be missed. The variety of adverse conditions to which malnourished children are exposed (Fig. 5.1) may have additive or interactive effects. For example, Porter and colleagues (1984) have shown that the weight of mice was influenced by infections and chemical agents when there was a moderate reduction in food and water availability.

Choice of Behavioral Measure

The previous section indicated that behavioral measures should not be restricted to those thought to reflect intellectual functioning. The first consideration in the choice of a behavioral measure should be to choose one that is homologous to the behavior of interest in humans, if such behaviors have been identified. If that information is not available, the choice of measures should be based upon prior knowledge of the likely metabolic and neurochemical effects of a nutrient. Behaviors should then be chosen that are known to be influenced by these metabolic or neurochemical effects. For example, Pardridge (1986) reasoned that the product of aspartame breakdown, phenylalanine, would compete with tryptophan for uptake into the brain. Thus, because brain levels of serotonin are a function of tryptophan availability (Wurtman and Fernstrom 1975), behaviors known to depend upon serotonin should be examined for effects of aspartame, e.g. seizures (which are more likely when levels of inhibitory neurotransmitters are decreased), sleep, and activity. In addition, the stage of development at which the nutrient is manipulated must be considered. Those systems known to be vulnerable during the developmental period in which the nutrition is manipulated should be considered candidates for behavioral measures. If there is a variety of behaviors which it is possible to study, the behavior with the better understood neural underpinnings is the obvious choice.

The sensitivity in detecting behavioral effects is increased if a multivariate approach to measuring behavior is taken. It is particularly important to measure the rate, intensity and timing of a behavior. These variables may be more sensitive than simple measurement of correctness of response. Alterations of circadian rhythms in variables such as activity or food intake can be sensitive indicators of behavioral effects of nutrients. In addition, as Thompson and colleagues discuss elsewhere in this volume (Chapter 6), complex tasks can be especially sensitive to nutrient effects. Thus, they describe the utility of repeated acquisition tasks, where chains of choices can be linked together to increase the complexity of the task.

When there is no information available to inform the choice of behavioral measures, the best approach may be to employ a panel of tests such as those used by behavioral teratologists (Adams 1986). Such screening batteries are useful because they assess the animals' ability to perform a variety of motor and sensory functions, as well as their learning ability.

Choice of Animal Model

As discussed above, the differences between humans and animals that make animal models so useful also lead to difficulties in generalization of findings from animals to humans. The optimal strategy is to choose a species with which one obtains the best balance between the similarities and differences. Unfortunately, researchers often choose animals with which they have always worked, i.e. the most convenient, without asking whether they respond to the nutrient as humans do. Investigators are usually sensitive to the fact that the metabolism of the nutrient must be similar in the human and animal model, but in addition, the sensory evaluation of the nutrient must be similar. For example, elsewhere in this volume (Chapter 11), Rolls and Hetherington state that rats do not perceive aspartame as sweet, and thus are not good animal models for studies of aspartame effects upon intake. In addition to determining whether the nutrient is handled similarly in the model species, similarity in the behavioral systems of interest must be ascertained. The best evidence for such similarity is that the neurologic substrate for the behavior is the same. Similar-looking behaviors cannot be assumed to have the same neural substrate nor serve the same functions for two different species. However, as noted above (p. 74) an animal behavior may be predictive of a human response even though the behavior is clearly not homologous to that in the human. Such predictive value must be empirically determined.

Conclusions

The study of the behavioral effects of nutrients requires an interdisciplinary approach. Often, however, this research is carried out by psychologists, who have little training in nutrition and metabolism, or nutrition scientists, who have little training in behavior. Because animal models and their interpretation are very complex, the investigator must have expertise in the study of animal behavior or, alternatively, collaborate with someone who has this expertise. The study of malnutrition perhaps best illustrates the need for an interdisciplinary approach. The literature is very difficult to summarize because of its contradictory findings. Its low reliability is due in part to the failure to understand the complexity of the undertaking, and the hubris of assuming one can study behavior or nutrition without adequate training. Thus, the quality of research in this field depends upon adequate interdisciplinary training and the establishment of collaborations between nutrition scientists and behavioral scientists.

Acknowledgement. Preparation of this paper was supported by Research Scientist Development Award No. MH00621 from the National Institute of Mental Health and HD04024 from the National Institute of Child Health and Human Development.

References

Adams J (1986) Methods in behavioral teratology. In: Riley EP, Vorhees CV (eds) Handbook of behavioral teratology. Plenum Press, New York, pp 67–100

Barrett DE (1987) Undernutrition and child behavior: what behaviors should we measure and how should we measure them? In: Dobbing J (ed) Early nutrition and later achievement. Academic Press, New York, pp 86–127

Barrett DE, Radke-Yarrow M (1985) Effects of nutritional supplementation on children's responses to novel, frustrating, and competitive situations. Am J Clin Nutr 42:102–120

Bhide PG, Bedi KS (1984) The effects of a lengthy period of environmental diversity on well-fed and previously undernourished rats. I. Neurons and glial cells. J Comp Neurol 227:296–304

Bunge R, Johnson M, Ross CD (1978) Nature and nurture in development of the autonomic neuron. Science 199:1409–1416

Chase HP, Crnic LS (1977) Undernutrition and human brain development. In: Mittler P (ed) Research to practice in mental retardation: biomedical aspects. Int Soc Sci Study Metal Defic, vol 3, pp 337–346

Chavez A, Martinez C (1975) Nutrition and development of children from poor rural areas. V. Nutrition and behavior development. Nutr Rep Int 11:477–489

Cravioto J, Delicardie ER (1979) Nutrition, mental development, and learning. In: Falkner F, Tanner JM (eds) Human growth, vol 3, Neurobiology and nutrition. Plenum Press, New York, pp 481–511

Crnic LS (1976) The effects of infantile undernutrition on adult learning in rats: methodological and design problems. Psychol Bull 83:715–728

Crnic LS (1980) Models of infantile malnutrition in rats: effects upon maternal behavior. Develop Psychobiol 13:615–628

Crnic LS (1984) Early experience effects: evidence for continuity? In: Emde RN, Harmon RJ (eds) Continuities and discontinuities in development. Plenum Press, New York, pp 355–368

Crnic LS, Reite ML, Shucard DW (1982) Animal models of human behavior. In: Emde RN, Harmon RJ (eds) The development of attachment and affiliative systems. Plenum Press, New York, pp 31–42

Delicardie ER, Cravioto J (1975) Language development in survivors of clinical severe malnutrition. Proc 9th Int Congr Nutr 1972, vol 2. Karger, Basel, pp 322–329

Denny MR, Ratner SC (1970) Comparative psychology. The Dorsey Press, Homewood, Illinois, p 3

Dobbing J (1968) The developing brain. In: Davison AN, Dobbing J (eds) Applied neurochemistry. FA Davis, Philadelphia, pp 287–316

Grantham-McGregor S (1987) Field studies in early nutrition and later achievement. In: Dobbing J (ed) Early nutrition and later achievement. Academic Press, New York, pp 128–174

Hinde RA (1976) The use of differences and similarities in comparative psychopathology. In: Serban G, Kling A (eds) Animal models of human psychobiology. Plenum Press, New York, pp 187–202

Im HS, Barnes RH, Levitsky DA (1976) Effect of early protein-energy malnutrition and environmental changes on cholinesterase activity of brain and adrenal glands of rats. J Nutr 106:342–349

Katz HB, Davies CA, Dobbing J (1982) Effects of undernutrition at different ages early in life and later environmental complexity on parameters of the cerebrum and hippocampus in rats. J. Nutr 112:1362–1368

Klein RE, Lester BM, Yarbrough C, Habicht J-P (1975) On malnutrition and mental development. Proc 9th Int Congr Nutr 1972, vol 2. Karger, Basel, pp 315–321

Lee TM, Moltz H (1984) The maternal pheromone and deoxycholic acid in the survival of preweanling rats. Physiol Behav 33:931–935

Lennenberg EH (1967) Biological foundations of language. Wiley, New York

Leon M, Moltz H (1972) Development of the pheromonal bond in the albino rat. Physiol Behav 8:683–686

Levine S, Wiener S (1976) A critical analysis of data on malnutrition and behavioral deficits. Adv Pediatr 22:113–136

Levitsky DA, Barnes RH (1972) Nutritional and environmental interactions in the behavioral development of the rat: long-term effects. Science 176:68–70

Levitsky DA, Barnes RH (1973) Malnutrition and animal behavior. In: Kallen DJ (ed) Nutrition, development and social behavior. US DHEW, Washington, DC, pp 3–16

Lynch A (1976) Passive avoidance behavior and response thresholds in adult male rats after early postnatal undernutrition. Physiol Behav 16:27–32

Massaro TF, Levitsky DA, Barnes RH (1974) Protein malnutrition in the rat: its effects on maternal behavior and pup development. Develop Psychobiol 7:551–561

McKay HK, Sinisterra L, McKay A, Gomez H, Lloreda P (1978) Improving cognitive ability in chronically deprived children. Science 200:270–278

Moltz H (1984) Of rats and infants and necrotizing enterocolitis. Perspect Biol Med 27:327–335

Mora JO, Herrera MG, Sellers SG, Ortiz N (1981) Nutrition, social environment and cognitive performance of disadvantaged Colombian children at three years. In: Nutrition in health and disease and international development. Symposia from the XII international congress of nutrition. Alan R. Liss, New York, pp 403–420

Morris R (1984) Developments of a water-maze procedure for studying spatial learning in the rat. J Neurosci Methods 11:47–60

Pardridge WM (1986) Potential effects of the dipeptide sweetener aspartame on the brain. In: Wurtman RJ, Wurtman JJ (eds) Nutrition and the brain, vol 7. Raven Press, New York, pp 199–240

Plaut SM (1970) Studies of undernutrition in the young rat: methodological considerations. Develop Psychobiol 3:157–167

Porter WP, Hindsdill R, Fairbrother A et al. Toxicant–disease–environment interactions associated with suppression of immune system, growth, and reproduction. Science 224:1014–1017

Richardson SA (1973) Brain and intelligence. National Education Press, Hyattsville, Maryland

Smart JL (1977) Early life malnutrition and later learning ability: a critical analysis. In: Oliverio A (ed) Genetics, environment and intelligence. Elsevier/North Holland, Amsterdam, pp 215–235

Smart JL (1980) Attempts at equivalent maternal care for well-fed and underfed offspring: are the problems appreciated? Develop Psychobiol 13:431–433

Smart JL (1984) Animal models of early malnutrition: advantages and limitations. In: Brozek J, Schurch B (eds) Malnutrition and behavior: critical assessment of key issues. Nestlé Foundation, Lausanne, pp 444–459

Smart JL (1987) The need for and relevance of animal studies of early undernutrition. In: Dobbing J (ed) Early nutrition and later achievement. Academic Press, New York, pp 50–87

Smart JL, Preece J (1973) Maternal behavior of undernourished mother rats. Anim Behav 21:55–56

Smart JL, Whatson TS, Dobbing J (1975) Thresholds of response to electric shock in previously undernourished rats. Br J Nutr 34:511–516

Strupp BJ, Levitsky DA (1983) Early brain insult and cognition: A comparison of malnutrition and hypothyroidism. Develop Psychobiol 16:535–549

Wiener SG, Fitzpatrick KM, Levin R, Smotherman WP, Levine S (1977) Alterations in the maternal behavior of malnourished rats. Develop Psychobiol 10:243–254

Wurtman RJ, Fernstrom JD (1975) Control of brain monoamine synthesis by plasma amino acids. Am J Clin Nutr 28:538–647

Chapter 6

Behavioral Procedures for Assessing Biological Variables: Implications for Nutrition Research

T. Thompson, C. Pollock and C. W. Gershenson

Nutrition, Learning and Behavior Problems of Children

Between 10% and 28% of school-aged children have learning disabilities (Schonhaut and Saltz 1983). These children are inattentive, hyperactive, and have poor memory and language skills; however, no specific behaviors unique to them have been identified (Kirk and Elkins 1975; Hallahan and Kauffman 1976; Norman and Zigmund 1980). Factors contributing to learning disabilities include psychosocial disadvantage (Tucker 1980), exposure to environmental toxins (Needleman et al. 1979) and maternal drug use during pregnancy (Streissguth et al. 1986). These insults often occur during periods of rapid central nervous system growth and are correlated with cognitive and behavioral sequelae. Nutritional deficiencies may also contribute to learning disabilities. For example, iron deficiency anemia among developing preschool children is associated with biochemical, enzymatic and morphologic abnormalities (Pollitt and Leibel 1976). Poor learning and memory have been associated with diets deficient in thiamin, folacin, zinc, copper, vitamin B6, and magnesium (Yehuda 1987; Levine and Krahn 1988). Experimental and clinical malnutrition have been associated with impaired mental function, delayed skeletal–muscular growth, and alterations in long-term brain composition (Latham 1974; Brozek 1978). Hypo- and hyperactivity, irritability, diminished attention span, reduced scores on early tests of cognitive development and sensory motor function have been reported in nutritionally deficient children.

The precise relation between degree and duration of nutritional deficiency and resulting behavioral and cognitive impairment is poorly understood. It might be expected that poor nutrition would also influence basic processes underlying children's learning. Children suffering from severe protein and energy deficiency exhibit low responsivity to the environment and poor attention maintenance. Children recovering from malnutrition show deficits in motor and cognitive development not limited to one area of intellectual function and these behavioral and attentional deficits may persist after nutritional rehabilitation (Pollitt and

Thomson 1977; Evans et al. 1980; Galler 1986). IQ scores of children with an attention deficit disorder may not initially differ from those of normal controls, but ultimately a significant difference is found (Douglas and Peters 1979). Thus, over the short run (several weeks or months) effects of nutrition on processes underlying learning might not be manifested in overall intellectual status, but a sustained deficiency could lead to a persistent cognitive dysfunction resulting in a cumulative learning disability (see Barrett 1987).

Traditional Measures Used in Diet and Behavior Research

Research on psychological effects of malnourishment in children has generally relied on standardized intelligence and achievement tests to assess the effects of malnutrition. While these norm-referenced tests (e.g., Bayley Scales of Infant Development, Stanford-Binet, WISC) are useful for evaluating the current status of a child relative to a population, they reveal little about the behavioral processes responsible for differences between groups. Intellectual performance of children with a history of malnutrition continues to be depressed into their adolescent years (Galler et al. 1986). Poor academic progress is attributed to deficits in classroom behavior, especially attention problems, and not reduced IQ (Galler et al. 1984). Academic achievement of children diagnosed as learning-disabled is often below that expected for their age and IQ. Observation of children who have suffered other forms of development insults (CNS trauma, lead exposure) indicates that the nature of their learning deficit is not manifest in a reduced IQ score. Since a cognitive deficit may coexist with an above average IQ score, more specific tests may be required to identify a subtle functional impairment that results in a learning disability and poor academic achievement (St. James-Roberts 1979; Bellinger and Needleman 1982; Rumsey and Rapoport 1983).

It has proved difficult to isolate the effect of nutritional variables on learning (Pollitt and Thomson 1977; Rumsey and Rapoport 1983). Malnutrition is frequently accompanied by other forms of social and economic deprivation, infectious disease, and adverse environmental circumstances that could also affect a child's intellectual development (Latham 1974). Data on the premorbid development of malnourished children are often unavailable and it is usually possible to obtain only an approximate determination of present nutritional status (Pollitt and Thomson 1977). Field studies have relied on group designs and employed standard achievement and intelligence tests to investigate the relation between poor nutrition and cognitive development. There are drawbacks to group designs. First, it may not be possible to find a homogeneous sample and appropriately matched controls; interpretation of results is problematic because of differences between groups. The control group may be from disadvantaged environmental circumstances and investigators may be analyzing poor performance on a cognitive test against a background of generally depressed performance. Secondly, data from group averages obscures individual response patterns and the interaction of different variables affecting children's responses (Baxley and LeBlanc 1976; Baumeister 1984; Spring 1986). This type of data does not represent the performance of any particular individual and does not identify any processes affecting a child's ability to learn.

Standard intelligence and achievement tests have not proved useful for identifying processes affecting learning. These measures describe current cognitive status (Strauss and Allred 1986). Their usefulness for short-term studies is limited because the standard error of measurement is large and the number of test items, sampling concepts and skills acquired during a short period is small (Gadow and Swanson 1986). Research in pediatric psychopharmacology has demonstrated that achievement tests are not sensitive to drug effects on learning, and that failure to document effects of a treatment variable may be due to insensitivity of the measure used (Gadow and Swanson 1986). Since it appears that nutritional deficits (e.g. iron deficiency) or excesses (e.g. lead) affect processes underlying learning, it is important to differentiate between *cognitive state variables* assessed by neuropsychological instruments and *psychologic process variables* revealed by laboratory tasks.

Very little nutritional research with human subjects has focused on the repeated measurement of changes in psychologic process variables *within subjects* over time. As a result, the present discussion will emphasize related research in which pharmacologic factors have been studied at various parameter values, within individual subjects. Limited research with non-human subjects using related methods will also be examined.

Processes that Influence Performance

Two major processes influence performance on learning tasks: *attention* and *short-term memory*. Both are critically important to normal development. *Attention* refers to the degree to which a specific aspect of a stimulus or cue controls the behavior of the subject stimulated. Children are required to respond selectively to letters, colors, numbers, features of spoken words and other sounds. Teachers commonly remark that children with learning problems in school have difficulty "paying attention", i.e., coming under the control of verbal or written instructions. It is extremely difficult, if not impossible, to study attentional processes under relatively uncontrolled classroom conditions. Laboratory procedures have been designed to assess basic features of attention believed to play a role in attending in educational and home settings.

The second psychologic process involved in learning is *short-term memory*. A child who is presented with an instruction or a piece of information and then asked to repeat the material after a short delay period is usually able to do so; however, if a longer delay is interpolated between presenting the instructional cue and the request to repeat it, the child may be unable to do so. Deficits in short-term memory are common in aging (e.g. Davis and Mumford 1984), in children with brain damage associated with exposure to toxins (Needleman et al. 1979) and in dyslexic children (Cohen 1982). Inability to retain information presented for even a few seconds can markedly interfere with learning; impairment in short-term memory is a possible mechanism underlying changes in learning ability.

Methods for Measuring Short-Term Memory and Attention

Delayed Matching-to-Sample

One method employed in investigating short-term memory is the delayed matching-to-sample procedure. The child is asked to select a correct matching stimulus (e.g. a figure of a tree) from an array of possible alternative cues (e.g. dog, house, tree) after having been presented with the sample cue (e.g. tree) several seconds earlier. A measure of memory is obtained by varying the time between sample cue presentation and the opportunity to select the correct matching cue (this interval is often called the *retention interval*). This procedure, developed by Cumming and Berryman (1961), has been used in numerous pigeon and primate studies with the retention interval varying from 0 to 8 seconds or longer; as the retention interval increases, the proportion of correct responses decreases (D'Amato 1972). The matching-to-sample procedure has also been used to study short-term memory in children (Over and Over 1967; Bryant 1969; Constantine and Sidman 1975; Williamson and McKenzie 1979; Sidman et al. 1982). Among the variables studied with this procedure have been the duration of the retention interval and the effect of presenting an interfering irrelevant cue during the delay interval. The results have been similar to those obtained with laboratory primates and are consistent with findings of standard clinical neuro-psychologic assessments of memory functions (see Peterson and Peterson 1959; Murdock 1962; Glanzer and Cunitz 1966).

A modification of the delayed matching procedure was developed by Cumming and Berryman (1965) in which the length of the delay is determined by the response in the previous trial. If the response is correct, the delay interval is increased; the delay interval is decreased following an incorrect response. The measure of short-term memory is the maximum interval at which correct responding occurs (called the *limits of delay*). Bakke (1986) modified the procedure to assess the reproduction of the sample positions: a titrated sequence position reproduction with a group of persons with Down's syndrome.

Sequence Reproduction Procedures

Procedures developed to study attention and short-term memory in laboratory settings have been successfully employed with learning disabled children. These procedures include disjunctive reaction time (e.g. Dykman et al. 1971), flanker task (e.g. Higgins and Turnure 1984), and sequence reproduction (e.g. Mackay 1975). While disjunctive reaction time and flanker task are suitable for school-aged and adult human subjects, they are not designed for use with very young children. The *sequence reproduction* method, however, has been widely used to study memory functions in primates and in both normal and retarded children and adults.(Sidman and Rosenberger 1967; Medin 1969; Mackay and Brown 1971; Brown et al. 1972; Mackay 1975). The subject is presented with a series of stimuli, typically illuminated buttons or panels, for increasingly short exposure intervals. After exposure to several illuminated buttons in sequence, the subject is asked to press the same buttons in the correct order. Initially, when each panel is illuminated for an interval substantially longer than the threshold for attention, the sequence is correctly reproduced without difficulty. However, as the duration for which each panel is illuminated is shortened, it becomes increasingly difficult for the subject to detect the presence of the stimulus, since it appears as a brief flash. People with retardation are more affected by the rate of presentation and the length of delay interval than are non-handicapped subjects (Ellis 1970). The findings are consistent with the results obtained using standard clinical procedures (Peterson and Peterson 1959) that manipulate the length of exposure or presentation time (Murdock 1960; Hellyer 1962). Errors in correct sequence reproduction as the duration of the illuminated stimuli grows shorter is a sensitive measure of attention in monkeys (D'Amato and Worsham 1972) and in pigeons (Roberts and Grant 1974).

As with the simpler matching-to-sample procedure, errors increase as the retention interval (Medin 1969) and the sequence length to be retained are increased. However, sequences as long as ten members have been taught to laboratory primates and children with retardation (Sidman and Rosenberger 1967; Mackay and Brown 1971; Mackay 1975). The sequence reproduction procedure is functionally similar to traditional instruments assessing memory; in these clinical tests of mental ability, memory is evaluated by interpolating longer delays between the time of presentation and testing (Murdock 1961). Similarly, in sequence reproduction tasks, human memory is also reduced by increasing the rates of stimulus presentation (Aaronson 1967).

Repeated Acquisition Procedures

Ultimately, we are concerned with the degree to which children's learning ability is altered by nutritional status. There have been a variety of definitions of *learning* proposed over the years. It is generally agreed that as learning progresses, fewer errors are made in executing a given performance (Catania 1984). The hallmark of early child development is an extraordinary rate of new learning; this is also the time that cognitive and behavioral dysfunctions may first become apparent (Strauss and Allred 1986). Golub and co-workers (1985) have studied a black–white discrimination reversal learning task in Rhesus monkey infants reared on a zinc deficient diet from birth. They found zinc deficient infants were impaired in reversal learning. In Rhesus monkeys subjected to severe zinc deficiency and tested at adolescence, Golub et al. (1988) found the zinc-deprived animals required 2–3 times as many trials for visual discrimination learning reversal.

Learning ability is one of the most difficult behavioral phenomena to study in a laboratory setting, especially if one is interested in assessing changes that occur in an individual's learning ability over time. Methods tracking progressive changes in learning ability for a given subject over days, weeks, and months have been developed over the past 25 years. Boren (1963) devised a learning task called the *repeated acquisition of behavioral chains* that required a subject to respond in a predetermined sequence on a specified number of operanda. This operant technique solved the problem of how to assess changes in learning ability within subjects: each session the subject was required to learn a new sequence of responses and the pattern of acquisition and number of errors reached a steady state from session to session. This steady state of relearning serves as a baseline to evaluate variables that affect acquisition.

The repeated acquisition procedure is similar to the approach used by Harlow (1959) to study learning sets in monkeys. Rhesus monkeys were presented with a long series of discrimination problems; in each problem they were presented with two objects, one of which had a reward underneath it. Once the animal learned which object was the correct choice, it could be successful on all subsequent trials since the same object was correct on each trial. The learning of initial problems happened slowly because the monkey had to discover by trial and error which object was correct, but performance improved as additional problems were presented. This "learning how to learn" requires coming under the control of specific cues (e.g. form not color) and was called the formation of learning sets.

The repeated acquisition of behavioral chains technique was developed to study acquisition within an individual subject (Boren and Devine 1968). Rhesus monkeys were trained to learn a new sequence of responses each session; once the nature of the task had been mastered, the numbers of errors in each session of new learning remained relatively constant. Subjects were presented with twelve levers arranged in four groups of three and trained to learn that for each session, the reinforced chain would be a new sequence of responses. At the start of the chain, the lights above the first group of three levers were illuminated, when the subject pressed the correct lever, the lights over the next group were lit. If the monkeys pressed an incorrect lever, the press resulted in a brief time-out (lights turned off). When a correct four-response sequence had been completed, reinforcement (a food pellet) was given.

Donald Thompson (1970) modified the repeated acquisition procedure for use with pigeons. The four groups of three response keys were differentiated by color instead of position to allow more flexibility in adjusting the difficulty of the learning task. When light cues were removed and birds had only position cues as the basis for correct responding, the number of errors increased. This demonstrated sensitivity and reversibility of the steady-state baseline indicating it would be useful for studying other variables that influence learning.

Subsequent studies have demonstrated dose-related disruptive effects of various drugs (e.g. cocaine, D-amphetamine, phenobarbital, chlorpromazine) on the learning baseline (Thompson 1974, 1980; Thompson and Moerschbaecher 1978, 1979 a,b, 1981; Schrot and Thomas 1983; Cleary et al. 1988). The procedure has been used to evaluate effects of drug combinations (Thompson 1980; Thompson and Moerschbaecher 1981) and has been modified to study drug effects on retention in monkeys (Thompson et al. 1986). Other investigators (Schrot et al. 1984; Paule and McMillan 1986) have used the procedure to study effects of exposure to toxic substances (e.g. carbon monoxide, trimethyltin) and other environmental insults (microwave exposure) on learning in both pigeons and rats. The effects of seizures, neuroleptics, and anticonvulsants on animal learning have also been assessed using the repeated acquisition procedure (Weinberger and Killam 1978; Picker and Poling 1984; Paule and Killam 1986; Poling et al. 1986; Delany and Poling 1987).

Winston and T. Thompson (unpublished data) modified these procedures for use with Rhesus monkeys. These animals were trained in the repeated acquisition task that required them to learn a similar, but slightly different, response

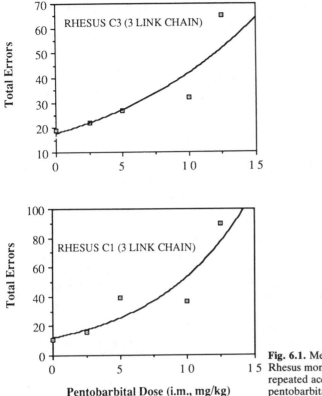

Fig. 6.1. Mean total errors for two Rhesus monkeys on a 3 link chain repeated acquisition task as a function of pentobarbital dose.

sequence during each daily test period. The task involved learning to press three illuminated panels sequentially, each in the presence of a different colored light. At the start of each session, the monkey must discover which of the three panel responses is correct in the presence of a given colored light (e.g. red). After correctly completing three responses in the presence of three different colors (a left panel press in the presence of red, far right press in the presence of blue, etc.) a food reinforcer is presented. Since the sequence of correct responses changed each day, the monkey had a new, but similar problem to learn. Figure 6.1 shows learning curves for two monkeys tested at a range of pentobarbital (intramuscular) dosages. There was an inverse relation between pentobarbital doses and ability to learn.

Use of Repeated Acquisition Procedures to Assess Biological Variables

The repeated acquisition procedure has also been used to assess drug effects in humans. Walsh and Burch (1979) tested the effects of aspirin, acetaminophen, caffeine, diphenhydramine, and dimenhydrinate on learning in three Navy divers. Error rates and time to complete the task increased after administration of each of the drugs. Fischman (1978) had adult volunteers learn a repeated acquisition task after cocaine and amphetamine administration. Response rates were unaffected by cocaine, regardless of the dose administered; however, the number of errors per test increased in a dose-related fashion. Amphetamine did not affect response rate and caused a decrease in the number of errors made while learning the sequence. Walker (1981) examined effects of orally administered amphetamine on learning by adult human subjects and found the number of errors decreased and response rates increased. Walker and Sprague (unpublished data) taught a repeated acquisition task to hyperkinetic children; then subjects received either orally administered methylphenidate hydrochloride or placebo. Methylphenidate administration resulted in a decrease in errors and an increase in response rate. Desjardins and co-workers (1982) investigated the effects of intravenous diazepam on the acquisition of response chains in adult humans. Diazepam decreased the overall rate of learning, and in two of three subjects, errors were increased in the learning condition at doses lower than those required to disrupt accuracy once the task had been learned.

The procedure has also been used to examine the effects of other biological variables on learning. Curley and Hawkins (1983) studied learning by six adult military personnel during a heat acclimatization regime; even after subjects exhibited physiological adaptation to the heat exposure, performance on the learning task was impaired. Perone and Baron (1982) used a repeated acquisition task to study learning by three younger and three older men. During paced sessions (i.e., responses had to occur within a specified interval) the behavior of the older men was disrupted, with increases in both omission and commission errors. The acquisition of new sequences was disrupted far more than performance of the established sequence.

Use of Repeated Acquisition Procedures to Assess Learning-Disabled People

Thompson and colleagues (1990) studied learning in young adults with moderate retardation using a repeated acquisition procedure. Many of the learning problems these people encounter in self-care (e.g. dressing, grooming), vocational and academic (e.g. learning telephone numbers) tasks involve chains of responses that must be emitted in a specific order in the presence of distinctive cues. Because of differences among tasks, it is difficult to study variables that influence such heterogeneous learning problems. The repeated acquisition procedure circumvents this difficulty and seems to be sensitive to at least two classes of variables that influence learning in other situations: chain length and consequence of errors.

The subject's task in the first part of this study was to press a series of panels in the presence of distinctive colors, for example: yellow – first panel correct; green – second correct; blue – third correct (i.e., a chain of 1–2–3). When the response chain was completed, the panel lights were extinguished for 1 second and an edible reinforcer was dispensed. When the chain length was increased, from 3 to 5 links, more initial errors were made, and a longer period was required to reach asymptotic performance. Moreover, as with learning of certain responses in clinical or educational settings, the consequences of errors influenced the degree of learning. If there was no "cost" associated with an error, learning was slow.

However, if a brief time-out was arranged following each link in the chain (i.e., the incorrect quadrant was selected) learning improved as the length of the time-out increased from 1 to 8 seconds. Thus, both task difficulty (i.e., chain length) and the consequences of making an incorrect choice influence learning rate; these factors are also operative in educational settings for children with learning problems.

Use of Repeated Acquisition Procedures to Assess Retardation and Dementia

The repeated acquisition procedure has been modified to evaluate effects of neuroleptic drugs on learning. Subjects in these studies are adults with moderate to severe mental retardation who are currently receiving neuroleptic drugs for management of behavior problems. Each chain link is associated with four two-dimensional pictures of familiar objects located in each quadrant of a computer monitor screen (e.g. flower, clock, broom and wagon). The subject's task each day is to learn which of the four objects in a given link is correct. An error leads to a brief time-out, as in the Thompson and co-workers (1989) study, while pressing the correct quadrant removes the four images, and four new images are presented (e.g. bed, faucet, chair and dish). The number of links in the chain is gradually increased until an asymptotic performance of 10%–15% error rate is reached by the end of the session. This baseline measure of learning ability is taken at the neuroleptic dosage level at which the subject enters the study. In the course of the protocol, neuroleptic dosage is changed systematically, including a placebo level for periods of 2 to 4 weeks per condition. Optimal learning performance at each dosage level during the final week of that condition is used as a measure of the relation between oral dosage (and in some cases, neuroleptic blood levels), rate of learning and asymptotic performance.

No single relationship has been observed between neuroleptic dosage level and learning measures with the population being studied. For some of the research subjects, there is a nearly linear relation between errors in the learning task and neuroleptic dosage. For others, however, there appears to be a curvilinear relation between neuroleptic dosage and learning: the most rapid rate of learning occurs at an intermediate dosage range, with rates in both the higher and lower dosage ranges approximately half that of the intermediate. Figure 6.2 shows sample learning curves for a 25-year-old man with mild retardation, who was receiving thioridazine for aggressive behavior.

The repeated acquisition procedure is currently being used to evaluate the effects of progressive neuropathology associated with dementia of the Alzheimer type. Gershenson and Thompson (1988) have compared learning errors made by patients being followed as part of a longitudinal Alzheimer disease project with errors made by their matched controls, who were not suffering from a dementing condition, during performance of a repeated acquisition task. Figure 6.3 shows a learning curve of a single subject (77-year-old man) with Alzheimer's disease and the learning curve of a control subject (69-year-old woman). Note that not only is the rate of learning (i.e., the speed with which the curve approaches asymptotic performance) lower for the Alzheimer patient, but that also the ultimate level correct consistently fails to reflect a total mastery.

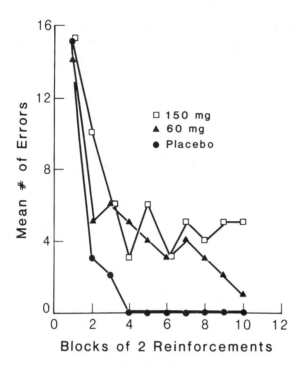

Fig. 6.2. Mean total within-session errors on the repeated acquisition learning task as a function of daily dosage of thioridazine. The subject is a 25-year-old man with retardation and serious behavior problems. Each curve is based on the data from the last 3 days at each dosage or placebo. Each data point is the average value for two completed chains (Thompson et al. 1986).

Advantages of Laboratory Tasks to Assess Learning

The use of a laboratory task (e.g. the repeated acquisition of behavioral chains) to assess changes in learning ability in an individual subject over time has several advantages compared with use of more face-valid learning tasks. The rate of learning different face-valid tasks over time cannot be quantitatively compared because of differences among tasks. One cannot quantitatively compare a child's ability to learn to tie his/her shoes, with the speed of learning to button or to do math problems. Moreover, laboratory tasks (such as sequence reproduction) can be modified systematically to increase or decrease complexity, permitting individualization of learning difficulty. This is important for studies in which young children of varying developmental and intellectual levels are involved. The repeated acquisition task is also repeatable. Once a subject has learned to tie his or her shoes, this task cannot be unlearned; the repeated acquisition procedure can measure learning in an ongoing manner and the learning of a new, but similar task each session allows for a repeated measure of learning.

There are also advantages to the individual subject design used in this procedure as opposed to a conventional independent group design. In a group design, averaging of data obscures results that may be applicable to an individual subject; it is difficult to identify the nature of mechanisms associated with the presence or absence of improvement (Gadow and Swanson 1986). In a single-subject design where each subject serves as his or her own control, individual performance is directly measured and findings are pertinent to the individual. The single-subject design is also useful for small sample sizes and in situations where variability among learning performance across subjects renders group

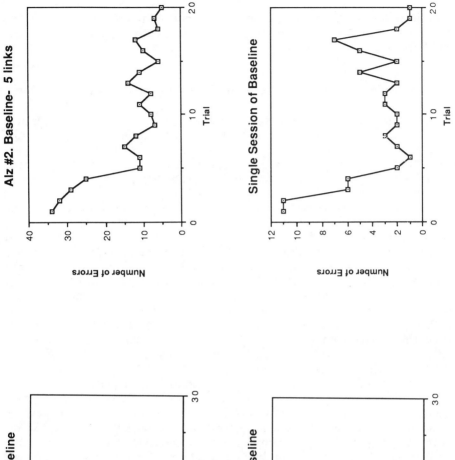

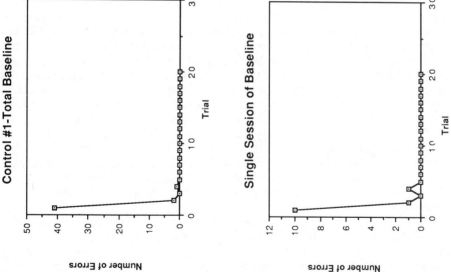

Fig. 6.3. Learning curves during the performance of a repeated acquisition task for a 77-year-old man with Alzheimer's disease and 69-year-old woman as a control. The top two graphs show the total number of errors across 20 trials on an 8 link (for control subject) and a 5 link (for Alzheimer subject) repeated acquisition task. Total errors are from the last 4 sessions of baseline taken over 2 testing days. The bottom two graphs (single session) comprise one of the four summed sessions of baseline.

statistics of little value. Within-group design eliminates intergroup variability and also takes into account the idiosyncratic tendencies of each individual (Baxley and LeBlanc 1976; Spring 1986).

Validity and Generality of Laboratory Measures of Learning

Despite these advantages in design and type of task employed, the question remains concerning the validity of the repeated acquisition procedure as a measure of learning functionally related to other learning required of people. Thompson and colleagues (1989) compared the rate of learning a sheltered workshop assembly task by young adults with moderate retardation (learning to assemble nuts, bolts, washers and related parts in the proper order) with the speed of learning the repeated acquisition laboratory task. They found the rank differences correlations ranged from 0.81 to 1.00 depending on the measure of learning used.

The repeated acquisition of behavioral chains procedure has been widely studied in the animal laboratory where it has been proved sensitive to pharmacologic manipulations; it has also been successfully adapted for use with human subjects to study variables that affect learning in an individual. However, the repeated acquisition of behavioral chains procedure has not been directly tested with a nutritional manipulation as the independent variable. Conclusions from both field studies of human malnutrition and laboratory models of animal malnutrition have reported that people and laboratory animals with nutritional deficiencies acquire skills and interact with their environments less effectively (Pollitt and Thomson 1977; Rogers et al. 1985).

The nature of research on the effects of nutritional deficits on learning in human subjects imposes ethical and methodologic limitations on experimental manipulation of the independent variable and prevents use of dependent measures such as growth rate or mortality. The influence of nutrition, like the effects of drugs and environmental toxins, on behavior are time- and dose-dependent (Connors and Blouin 1983); procedures have been developed for use in the animal laboratory to study the relationship between these substances and behavior. Behavioral preparations have been used in the animal laboratory to determine time and dose parameters of drug and toxin effects, to assess abuse potential of various drugs, and to investigate effects of these substances on auditory and visual discrimination and learning ability. Behavioral tests are nondestructive and the same organism can be assessed before, during and after exposure to a nutritional manipulation (Evans and Daniel 1984). Evidence indicates that behavioral measures developed for use in the animal laboratory can also be useful for evaluating variables that affect processes such as attention and short-term memory in human subjects.

In view of (1) the problems posed by attempts to measure process variables underlying learning in children in applied settings, (2) the validity of the repeated acquisition procedure as a measure of learning ability, (3) the proven sensitivity of the procedure to known biological independent variables (e.g. drug dosage, ambient temperature, neuropathologic condition), and (4) the range of animal and human subjects that have been evaluated using this procedure, it seems that the time is propitious for the repeated acquisition procedure to be evaluated as a tool for better understanding the relation between nutritional variables, learning ability and associated cognitive processes in children.

References

Aaronson D (1967) Temporal factors in perception and short-term memory. Psychol Bull 67:130–144

Bakke B (1986) The effects of delay and difficulty on a visual recall task with adults who have Down's syndrome. Unpublished PhD dissertation, Georgia State University, Atlanta

Barrett DE (1987) Undernutrition and child behavior: what behaviours should we measure and how should we measure them? In: Dobbing J (ed) Early nutrition and later achievement. Academic Press, New York, pp 86–119

Baumeister AA (1984) Some methodological and conceptual issues in the study of cognitive processes with retarded people. In: Brookes PH, Sperber R, McCauley C (eds) Learning and cognition in the mentally retarded. Lawrence Erlbaum, New Jersey

Baxley GB, LeBlanc JM (1976) The hyperactive child: characteristics, treatment, and evaluation of research design. Adv Child Dev Behav 11:1–35

Bellinger DC, Needleman HL (1982) Low level lead exposure and psychological deficit in children. Adv Dev Behav Pediatr 3:1–49

Boren JJ (1963) Repeated acquisition of new behavioral chains. Am Psychol 17:421

Boren JJ, Devine DD (1968) The repeated acquisition of behavioral chains. J Exp Anal Behav 11:651–660

Brown SM, Renew FC, Mackay HA (1972) Reproduction of sequences by monkeys: disruption of performance with 2 member samples after training with one 3 member sample sequence. Paper presented at the meeting of the Canadian Psychological Association, Montreal

Brozek J (1978) Nutrition, malnutrition, and behavior. Ann Rev Psychol 29:157–177

Bryant PE (1969) Perception and memory of the orientation of visually presented lines by children. Nature 224:1331–1332

Catania AC (1984) Learning, 2nd edn. Prentice-Hall, Englewood Cliffs

Cleary J, Ho B, Nader M, Thompson T (1988) Effects of buprenorphine, methadone and naloxone on acquisition of behavioral chains. J Pharmacol Exp Ther 247(2):569–575

Cohen RL (1982) Individual differences in short-term memory. In: Ellis NR (ed) International review of research in mental retardation, vol 11. Academic Press, New York, pp 43–47

Connors CK, Blouin AG (1983) Nutritional effects on behavior of children. J Psychiatr Res 17:193–173

Constantine B, Sidman M (1975) The role of naming in delayed matching to sample. Am J Ment Defic 79:680–689

Cumming WW, Berryman R (1961) Some data on matching behavior in the pigeon. J Exp Anal Behav 4:281–288

Cumming WW, Berryman R (1965) The complex discriminated operant: studies of matching to sample and related problems. In: Mostofsley DI (ed) Stimulus generalization. Stanford University Press, Stanford

Curley MD, Hawkins RN (1983) Cognitive performance during a heat acclimatization regimen. Aviat Space Environ Med 54:709–713

D'Amato MR (1972) Delayed matching and short-term memory in monkeys. In: Bower G (ed) The psychology of learning and motivation, vol 7. Academic Press, New York, pp 1–32

D'Amato MR, Worsham RW (1972) Delayed matching in the capuchin monkey with brief sample durations. Learn Motiv 3:304–312

Davis PE, Mumford SJ (1984) Cued recall and the nature of the memory disorder in dementia. Br J Psychiatr 144:383–386

Delaney D, Poling A (1987) Effects of methsuximide and mephenytoin on the behavior of pigeons under a repeated acquisition procedure. Pharmacol Biochem Behav 28:483–488

Desjardins PJ, Moerschbaecher JM, Thompson DM, Thomas JR (1982) Intravenous diazepam in humans: effects on acquisition and performance of response chains. Pharmacol Biochem Behav 17:1055–1059

Douglas VI, Peters KG (1979) Toward a clearer definition of the attentional deficit of hyperactive children. In: Hale GA, Lewis M (eds) Attention and cognitive development. Plenum Press, New York, pp 173–247

Dykman RA, Ackerman P, Clements S, Peters JE (1971) Specific learning disabilities: an attentional deficit syndrome. In: Myklebust HR (ed) Progress in learning disabilities, vol 2. Grune and Stratton, New York, pp 56–93

Ellis NR (1970). Memory processes in retardates and normals: theoretical and empirical considerations. In: Ellis NR (ed) International review of research in mental retardation, vol 4. Academic Press, New York, pp 1–32

Evans D, Hansen JDL, Moodie AD, van der Spuy HIJ (1980) Intellectual development and nutrition. J Pediatr 97:358–363

Evans HL, Daniel SA (1984) Discriminative behavior as an index of toxicity. In: Thompson T, Dews P, Barrett J (eds) Advances in behavioral pharmacology, vol 4. Academic Press, New York

Fischman MW (1978) Cocaine and amphetamine effects on repeated acquisition in humans. University of Chicago, Chicago (from Higher Cerebral Function and Behavior, abstract 2138)

Gadow KD, Swanson HL (1986) Assessing drug effects on academic performance. Adv Learn Behav Dis Suppl 1:247–279

Galler JR (1986) Malnutrition – a neglected cause of learning failure. Postgrad Med 80:225–229

Galler JR, Ramsey F, Solimano G (1984) The influence of early malnutrition on subsequent behavioral development. III. Learning disabilities as a sequel to malnutrition. Pediatr Res 18:309–313

Galler JR, Ramsey F, Forde V (1986) A follow-up study of the influence of early malnutrition on subsequent development. IV. Intellectual performance during adolescence. Nutr Behav 3:211–222

Gershenson C, Thompson T (1988) Laboratory studies of attention, memory, and learning in Alzheimer's disease patients and normal adults. Gerontologist [Special issue] 28:194A

Glanzer M, Cunitz A (1966) Two storage mechanisms in free recall. J Learn Verbal Behav 5:531–560

Golub MS, Gershwin ME, Hurley LS (1985) Studies of marginal zinc deprivation in rhesus monkeys: infant behavior. Clin Nutr 42:1229–1239

Golub MS, Gershwin ME, Hurley LS, Hendricks AG (1988) Studies of marginal zinc deprivation in rhesus monkeys. VIII. Effects in early adolescence. Am J Clin Nutr 42:1229–1239

Hallahan DP, Kauffman JM (1976). Introduction to learning disabilities: a psycho-behavioral approach. Prentice-Hall, Englewood Cliffs

Harlow H (1959) Learning set and error factor theory. In: Koch S (ed) Psychology: a study of a science, vol 2. McGraw-Hill, New York, pp 492–537

Hellyer S (1962) Frequency of stimulus presentation and short-term decrement in recall. J Exp Psychol 64:650

Higgins AJ, Turnure JE (1984) Distractibility and concentration of attention in children's development. Child Dev 55:1799–1810

Kirk SA, Elkins J (1975) Characteristics of children enrolled in the Child Service Demonstration Centers. J Learn Dis 8:630–637

Latham MC (1974) Protein-calorie malnutrition in children and its relation to psychological development and behavior. Physiol Rev 54:541–565

Levine AS, Krahn DD (1988) Food and behavior. In: Morley JE, Sterman MB, Walsh JH (eds) Nutritional modulation of neural function. Academic Press, New York, pp 233–247

Mackay HA (1975) Delayed matching of serial position sequences by retarded subjects: effects of sequence length. Paper presented at the meeting of the Eastern Psychological Association, New York, April 1975

Mackay HA, Brown SM (1971) Teaching serial position sequences to monkeys with a delayed matching-to-sample procedure. J Exp Anal Behav 15:335–345

Medin DL (1969) Form perception and pattern reproduction in monkeys. J Comp Physiol Pathol 68:412–419

Murdock BB (1960) The immediate retention of unrelated words. J Exp Psychol 60:222–234

Murdock BB (1961) The retention of individual items. J Exp Psychol 62:618–625

Murdock BB (1962) The serial position effect in free recall. J Exp Psychol 64:482–488

Needleman JL, Gunnoe C, Leviton A et al. (1979). Deficits in psychologic and classroom performance of children with elevated dentine lead levels. N Engl J Med 300:689–695

Norman CA, Zigmund N (1980) Characterisics of children labelled and served as learning disabled in school systems affiliated with Child Service Demonstration Centers. J Learn Dis 13:542–547

Over R, Over J (1967) Detection and recognition of mirror-image obliques by young children. J Comp Physiol Psychol 64:467–470

Paule MG, Killam EK (1986) Behavioral toxicity of chronic ethosuximide and sodium valproate treatment in the epileptic baboon. *Papio papio*. J Pharmacol Exp Ther 238(1):32–38

Paule MG, McMillan DE (1986) Effects of trimethyltin on incremental repeated acquisition (learning) in the rat. Neurobehav Toxicol Teratol 8:245–253

Perone M, Baron A (1982) Age-related effects of pacing on acquisition and performance of response sequences: an operant analysis. J Gerontol 37(4):443–449

Peterson LR, Peterson MJ (1959) Short-term retention of individual verbal items. J Exp Psychol 58:193–198

Picker M, Poling A (1984) Effects of anticonvulsants on learning: performance of pigeons under a repeated acquisition procedure when exposed to phenobarbital, clonazepam, valproic acid, ethosuximide and phenytoin. J Pharmacol Exp Ther 230(2)307–316

Poling A, Blakely E, White W, Picker M (1986) Chronic effects of clonazepam, phenytoin, ethosuximide, and valproic acid on learning in pigeons as assayed by a repeated acquisition procedure. Pharmacol Biochem Behav 24:1583–1586

Pollitt E, Leibel G (1976) Iron deficiency and behavior. J Pediatr 88(3):372

Pollitt E, Thomson C (1977) Protein-calorie malnutrition and behavior: a view from psychology. Nutr Brain 2:261–306

Roberts WA, Grant DS (1974) Short-term memory in the pigeon with presentation time precisely controlled. Learn Motiv 5:393–408

Rogers PJ, Tonkiss J, Smart JL (1985) Incidental learning is impaired during early-life undernutrition. Develop Psychobiol 19:113–124

Rumsey JM, Rapoport JL (1983) Assessing behavioral and cognitive effects of diet in pediatric populations. Nutr Brain 6:101–161

Schonhaut S, Saltz P (1983) Prognosis for children with learning disabilities. A review of follow up studies. In: Rutter M (ed) Developmental neuropsychiatry. Churchill Livingstone, Edinburgh, pp 542–563

Schrot J, Thomas JR (1983) Alteration of response patterning by D-amphetamine on repeated acquisition in rats. Pharmacol Biochem Behav 18:529–534

Schrot J, Thomas JR, Robertson RF (1984) Temporal changes in repeated acquisition behavior after carbon monoxide exposure. Neurobehav Toxicol Teratol 6:23–28

Sidman M, Rosenberger PB (1967) Several methods for teaching serial position sequences to monkeys. J Exp Anal Behav 10:467–478

Sidman M, Ruzin R, Lazer R, Cunningham S, Tailby W, Carrigan P (1982) A search for symmetry in the conditional discrimination of rhesus monkeys, baboons, and children. J Exp Anal Behav 37:23–44

Spring B (1986) Effects of foods and nutrients on the behavior of normal individuals. Nutr Brain 7:1–47

St. James-Roberts I (1979) Neurological plasticity and recovery from brain insult. Adv Child Dev Behav 14:253–319

Strauss JE, Allred LJ (1986) Methodological issues in detecting specific long-term consequences of perinatal drug exposure. Neurobehav Toxicol Teratol 8:369–373

Streissguth AP, Barr HM, Martin DC, Darby BL (1986) The fetal alcohol syndrome as a model for the study of behavioral technology of alcohol. In: Krasnegor NA, Gray DB, Thompson T (eds) Developmental behavioral pharmacology. Lawrence Erlbaum, New Jersey

Thompson DM (1970). Repeated acquisition as a behavioral baseline. Psychon Sci 21:156–157

Thompson DM (1974) Repeated acquisition of behavioral chains under chronic drug conditions. Pharmacol Exp Therap 188:700

Thompson DM (1980) Selective antagonism of the rate-decreasing effect of D-amphetamine by chlorpromazine in a repeated-acquisition task. J Exp Anal Behav 34:87–92

Thompson DM, Moerschbaecher JM (1978) Operant methodology in the study of learning. Environ Health Perspect 26:77–87

Thompson DM, Moerschbaecher JM (1979a) An experimental analysis of the effects of D-amphetamine and cocaine on the acquisition and performance of response chains in monkeys. J Exp Anal Behav 32:433–444

Thompson DM, Moerschbaecher JM (1979b) Drug effects on repeated acquisition. In: Thompson T, Dews PB (eds) Advances in behavioral pharmacology, vol 2. Academic Press, New York

Thompson DM, Moerschbaecher JM (1981) Selective antagonism of the error-increasing effect of morphine by naloxone in a repeated-acquisition task. J Exp Anal Behav 36:371–380

Thompson DM, Mastropaolo J, Winsauer PJ, Moerschbaecher JM (1986) Repeated acquisition and delayed performance as a baseline to assess drug effects on retention in monkeys. Pharmacol Biochem Behav 25:201–207

Thompson T, Suessbrick A, Gershenson CW, Joseph B (1990) A within-subjects procedure for studying learning by mentally retarded adults. (Submitted for publication)

Tucker JA (1980) Ethnic proportions in classes for the learning disabled: issues in nonbiased assessment. J Special Edu 14:93–105

Walker MK (1981) Effects of dextroamphetamine sulfate on repeated acquisition behavior and mood in humans: a preliminary report. Psychol Record 31:29–41

Walker MK, Sprague RE (1981) Effects of methlyphenidate hydrochloride on a repeated acquisition behavior in hyperkinetic children (unpublished)

Walsh JM, Burch LS (1979) The acute effects of commonly used drugs on human performance in hyperbaric air. Undersea Biomed Res 6:348–351

Weinberger SB, Killam EK (1978) Alterations in learning performance in the seizure-prone baboon: effects of elicited seizures and chronic treatment with diazepam and phenobarbital. Epilepsia 19:301–316

Williamson AM, McKenzie BE (1979) Children's discrimination of oblique lines. J Exp Child Psychol 27:533–543

Yehuda S (1987) Nutrients, brain biochemistry, and behavior: a possible role for the neuronal membrane. Int J Neurosc 35:21–36

Discussion

P. B. Dews

Methods of measuring behavioral response must take into account the type of effect under consideration, because time courses differ and experimental design must be matched to the question addressed. Some general principles apply to most methods appropriate to the study of effects of diet on behavior: these will be considered first, followed by discussion of particular types of effects of diet on behavior under the headings nutritional, pharmacologic and psychologic. The effects of behavior on diet will then be considered under the same three headings; this will be followed by concluding speculations.

General Requirements of Methods

The effects of diet on behavior arise from the chemical composition of the diet, the absolute and relative amounts of its components, temporal patterns of intake, and a whole variety of factors related to environmental circumstances before, during and after ingestion. All these factors can be objectively and quantitatively assessed; the list is not exhaustive, but the other factors can also be satisfactorily assessed. Behavior affects the environment and so can also be measured objectively and quantitatively. It should not be (but is) necessary to emphasize that methods of measuring the behavioral response to diet should be as objective and quantitative as the methods of assessing the dietary factors. The latter methods determine the chemical composition of dietary components and the amounts of the various ingredients in unequivocal quantitative terms. The former methods must permit the relevant behavioral components to be determined and their amounts and frequencies. As pencil marks on paper or keystrokes on a keyboard are objective phenomena and can be quantified, these requirements do not limit the range of possible behavioral effects that can be assessed.

The reason why objective and quantitative methods of measuring the behavioral response must be emphasized is because there is an atavistic concern in contemporary psychology with non-material, inferred factors such as cognition. Cognition is incommensurate with dietary variables and useless for illuminating effects of diet on behavior (or effects of behavior on diet). It has been suggested, for example, that "as the subject of behavior has matured, it became clear that we cannot understand actions without understanding mental processes and feelings." But there are no methods for studying mental matters directly. Verbal reports, questionnaires and interviews are available but such methods do not provide direct information on mental "processes." All the information that is available is contained in the answers to the questions which can be handled objectively. Fallacies arise in the "interpretation" of the verbal behavior itself. Introduction of tautologic or gratuitous intervening variables that do not derive from the objective results do not add to the amount of scientific information available, but may obscure real relationships.

Most studies of effects of diet on behavior are experimental studies. The experimenter sets the values of the dietary factors and decides which factors will be systematically changed between subjects or during the experiment and what are appropriate controls. The experimental design predetermines the analysis of results, which should be straightforward and direct. Biologically important effects should be clear and convincing from simple graphical presentations if the experimenter is going to accept them as significant, valid findings. Before announcing findings, however, the experimenter usually performs a statistical analysis of the results, because sometimes what appear to be clear effects could have arisen by chance with unacceptable frequency, more than 1 in 20, for example, or whatever odds are judged to be unacceptable. Usually, simple parametric statistics based on means and standard deviations will produce estimated odds. Transformations based on complex mathematical assumptions are rarely helpful. The exception is the logarithmic transformation of dose which has been overwhelmingly validated in pharmacology. Statistics are used in this type of experimentation to warn the experimenter when his results do not permit him to come to a clear conclusion on the relations under examination. Statistics here cannot be used to *establish* the reality of a relationship if the results themselves are not convincing. No matter how low the "*P*" value, all that can be said is that the results are too improbable to be due to chance, but not that the non-chance influence is the factor the study was designed to assess. Only the results themselves can do that. In experimental studies of this kind, therefore, statistical tests can make apparent relationships unconvincing but cannot, in themselves, establish significant relationships.

It must be emphasized that the foregoing remarks apply to the use of statistics in the particular types of experimental studies that are commonly used in nutrition and other physiologic sciences, and in experimental psychology. In other types of scientific study other types of statistics are needed. In some branches of biology, for example population genetics, the critical evidence is statistical in nature. These strictures are not against the use of statistics in science; they concern the use of statistics inappropriate to the type of investigation.

While the calculus of probability is invariant, applications to help in scientific decisions are as varied as the experimental designs and types of information that are generated in scientific studies. That is one reason why commercial packages of computer software for statistical tests are dangerous. Individual researchers buy and become familiar with the logistics of a particular program and then tend to use it whenever they have results to analyze. Unfortunately, that the numbers generated in a study can be entered into a statistical routine is no guarantee, indeed not even an indication, that the routine is appropriate for the proper handling of the results. So, for example, agricultural statistics are sometimes applied to experimental results in the physiologic and psychologic sciences where the field is not divided into plots and when a few hours, rather than a whole growing season, suffices to obtain a result or a replication. Purchasers of such packages should be required to pass a searching examination before being allowed to use them, in the same way that individuals are examined before being allowed to operate a dangerous instrument like an automobile or before being allowed to prescribe dangerous drugs.

One class of effect of diet on behavior that is of continuing concern and interest is the potentially permanent consequences on subsequent behavioral capabilities of extreme manipulations of diet during development. Assessment of such consequences requires long-term prospective studies with repeated assessments of individuals. The statistical handling of results of such studies is entirely different from the handling of results of acute experiments discussed so far. Analysis of variance is as inappropriate here as in acute experiments. Powerful techniques have been developed for prospective clinical studies that involve repeated assessments of individual subjects. The methods can be applied to laboratory clinical studies on long-term effects of diet prenatally and in infancy. Such methods do not seem to be as widely exploited as they should be.

Dietary effects on behavior as a branch of pharmacology have been discussed previously (Dews 1986). The rules that have been found to be necessary for sound conclusions in pharmacology need to be obeyed. The importance of systematic exploration of dose–effect relations, for example, has been stressed by others in this volume. Equally important is attention to details of the diet beyond the specific ingredient under study and to the nutritional status of subjects. Pharmacologists have traditionally paid little heed to the diets of their subjects, being content to leave their subjects to the routine of the laboratory. Such neglect is impermissible in studies of effects of diet on behavior.

Dietary effects on behavior have features that set them apart from most other pharmacological effects. Effects tend to be small. Usually in pharmacology, large effects can be produced by sufficiently increasing the dose (hence the old pharmacologic adage: enough of anything does something to more or less anything that is being measured). The rule does not generally apply to effects of diet on behavior. Even gross overintakes of many, indeed most, individual dietary components in non-deficient subjects are without detectable behavioral effects. Only when the component is constituting such a large part of the diet that balanced nutrition is compromised are effects seen. Some ingredients, such as vitamins A or D, can cause toxic effects when taken in great excess, but effects on behavior are only secondary. The intensity of the effects of even the components of the diet that have direct effects on behavior usually reaches an asymptote at which the effects of amounts in the dietary (as opposed to the pharmacologic) range are still modest. The restricted range over which effect varies with dose makes the determination of dose–effect relations in diet and behavior generally more difficult than in most of the rest of pharmacology. At the very least, however, it should be possible to identify levels so low that effects cannot be measured and the upper asymptote of intensity of effects.

What is meant by saying that, ordinarily, effects of diet on behavior are small? The behavioral effects of starvation may indeed be profound, but only when the *physiologic* effects of the starvation are profound and adequate to account indirectly for the behavioral effects. For malnutrition, even during development, to produce behavioral sequelae, it has to be extreme, as Crnic has described in rats, and as Cravioto and Delicardie (1979), for example, have described in humans. To suppose that because extremes of dietary manipulations can produce large effects, then small variations in diet will also produce effects, albeit tiny, is a fallacy. As levels of malnutrition become less extreme, effects on behavioral development asymptote towards zero. Malnutrition has to become more extreme than occurs in the USA, except rarely and sporadically, to cause permanent effects.

Studies of the effect of extreme malnutrition on development are of neurobiologic interest, because they help elucidate controlling mechanisms during development. The interest is in mechanism rather than in direct application to human circumstances. Selective effects of diet on behavior are usually changes of a few percent on the full scale of the behavioral responses being measured. They are small in relation to effects produced, for example, by many drugs or environmental changes. But even small effects are important, if they occur frequently in large numbers of people.

Effects of Diet on Behavior

Nutritional Effects

Nutritional effects on behavior occur when the diet lacks adequate amounts of required ingredients. Effects of extreme malnutrition are mentioned above. Evidence for behavioral effects of the range of diets consumed in the USA is surprisingly lacking. Perhaps, though, one should not be surprised. The integrity of the nervous system is strongly conserved in the face of malnutrition. For example, even during development, malnutrition sufficient to lower body weight by 80% reduced brain weight by only 18% (Dobbing and Widdowson 1965). When development is complete, effects of even extreme malnutrition are probably completely reversible. The behavioral effects of life-threatening diseases such as beriberi are reversed when the deficiency is corrected. Crnic confirms this conclusion, and further, has shown that even under conditions sufficiently severe to cause reduction in brain weight, *behavioral* changes are less than devastating. The changes are certainly real, but they are not greater than can be produced by environmental changes that are not as life-threatening as extreme malnutrition. A study of a Dutch population subject to severe famine for a well-defined period late in World War II could not detect effects on the subsequent school performance. The brain and its functions seem to be no more sensitive to effects of malnutrition than other organs and tissues of the body. It may be argued that brain changes are of more concern because they are by their nature less reversible than changes in other tissues. An atrophied liver can quite quickly grow back to normal, none the worse for its experience, but a brain that is malnourished at particular periods can never subsequently regain full normality. Has this been shown convincingly for malnutrition that is less than life-threatening?

It is established that children who have breakfast do better in school than children who do not have breakfast. It is not clear to what extent the effects of breakfast are determined by the specific nutritional properties of the components of the breakfast and to what extent the effects are non-specific effects of being less hungry. However, the distinction may not be important for practical purposes.

Pharmacologic Effects

In addition to nutritional effects of the diet – provision of energy and replenishment of the pools of substances that are needed to maintain the dynamic integrity of the body – there are dietary components with pharmacologic effects. Some, such as alkaloids and amines, are intrinsic components of the foodstuffs; others, such as pesticides, are residues; yet others, like aflatoxin, arise during storage; finally some, like preservatives, are deliberately added. Identification of behavioral effects of such ingredients is the concern of the emerging disciplines of behavioral pharmacology and toxicology, rather than of nutrition. Selective pharmacologic effects of a high carbohydrate or protein meal, or of ingestion of a single amino acid or oligopeptide is the concern of "diet and behavior." Much of the discussion of methods in this volume is about how to study effects of this kind. Pharmacologic effects differ from nutritional effects in usually lasting only as long as the concentration of a body component is changed by the diet. As noted, the nutritional effect of a dietary component will increase in magnitude with increasing dose of agent only until the supply of the component is adequate for an optimum diet. Pharmacologic effects may continue to increase with increasing dose over a broader range than for nutritional effects, though to a lesser extent than for real drugs. The distinction should not be overemphasized as there is little hard quantitative information.

Even the pharmacologic effects of normal dietary components are typically subtle. Until a few years ago the majority of pharmacologists were skeptical, if not of the reality of effects, at least of whether they were sufficient in magnitude to be worthy of concern. Workers such as Wurtman (1983) and Anderson (Hrboticky et al. 1985) have now proved that real effects occur at dietetically relevant doses. What are the effects? Anderson has studied mostly effects on behavior related to eating, both eating itself and written verbal behavior relevant to eating – surely reasonable because eating is an especially important behavior in relation to diet. Effects on sleepiness following tryptophan have been reported by more than one group. A variety of other effects have been described (Wurtman and Lieberman 1983). Almost all the effects are small. But, again, even small effects if they occur often in almost everybody have in aggregate a large societal impact.

Strong preference should be given to measuring effects directly in objective, quantitative terms yielding unambiguous numbers that can be plotted on a graph. Dichotomous (all or none) or categorical or other types of dependent variables are much less satisfactory. Crnic comments "particularly important is to measure the rate, intensity and timing of behavior." That is good advice. "These variables may be more sensitive than simple measurement of correctness of response," to which may be added that such measures will not only be more sensitive, they will be more *informative*.

It is necessary to explore dose–effect relations because almost nothing can be said quantitatively about the effects of an agent when only a single concentration or exposure has been studied. Indeed, it is hazardous to make even qualitative statements because not unusually even qualitative effects change with concentration. Some effects, for example, become manifest only when concentrations are above certain levels. There has been increasing recognition of the importance of dose–effect relations in the study of acute effects of agents down the years, but there is less recognition that dose–effect-type relations are equally important in long-term experiments. Also, such relations must be established when the independent variable being manipulated is not the amount of a particular agent, but the amount of food; or the amount of experience of a particular type, for example sensory; or the amount of exercise, and so on. In such studies, as in pharmacology, determinations only at single values can be misleading. An effect that changes in a systematic and orderly manner with change in independent variables is likely to be an effect of the agent rather than of a concomitant variable. Contrast this situation with a dichotomous experiment, e.g. a comparison of malnourished with well-nourished subjects. As Crnic points out, it is impossible to produce malnutrition by a discrete manipulation, leaving everything else unchanged. All the features that change with induction of malnutrition then become concomitant variables. What needs to be demonstrated is that it is not a concomitant variable, rather than the malnutrition itself, that is responsible for the measured changes. If a systematic series of quantitative grades of malnutrition could be imposed then changes that are systematically related to grade of malnutrition are more likely to be due to the malnutrition, because concomitant variables rarely co-vary exactly with the independent variable over a substantial range. Study of dietary effects on behavior is much more like behavioral pharmacology than it is like any other science, and the rules that have evolved, agonizingly, in pharmacology are applicable to the study of behavioral effects of agents when the agents are dietary components rather than conventional "drugs." Students of diet and behavior should not waste their time re-inventing the wheel.

Pharmacodynamic effects typically have a characteristic time course. An agent is ingested, concentrations in the body rise progressively over a period of time, effects intensify, and then the concentrations start to fall and effects abate. Obviously, measuring effects at an arbitrary point in time may give misleading results. In fact, the importance of measuring the time course of an effect is widely recognized and observed. Perhaps not adequately recognized, however, are the advantages of methods that permit the continuous assessment of behavior for hours, or for hours a day in chronic experiments, without the disruptive intrusion of the experimenter, change of environment and starting and stopping of observations. Such continuous assessments permit, in Thompson's words, the following of "process" and not just measurement at an arbitrary point in time.

Psychologic Effects

As discussed in earlier chapters on the origins of food preference, diet can have effects on behavior that are neither nutritional nor pharmacologic, as defined above. The behavior of seeking fine food, as in people turning out in severe winter weather to drive to a favorite restaurant, is certainly strong behavior related to diet, but has usually little relation to the nutritional or pharmacologic effects in the food served in that restaurant. Similarly, the behavior of paying large amounts of money for a fine wine is not dependent on the pharmacology of organism that one would expect that it would be relatively well protected from environmental insult." Indeed, learning, far from being a wonderful "higher" human attribute, is a primitive function, one that evolved hundreds of millions of years ago, at the time when cell membranes and intracellular cytostructure were evolving. The basic mechanisms of learning may have persisted in their essentials in the same way that the mechanism of hereditary transmission of information by DNA evolved and has persisted. Many people have studied the shaping of behavior, learning if you will, in invertebrates (see, for example, Dews 1959). The changes in behavior take place in a manner quite similar to how they take place in pigeon, monkey or human. Maybe they are truly homologous. The line of evolution of mollusks, that gave rise, for example, to the octopus, diverged from that which led to primates some 600 million years ago. If learning is a primitive cellular function, perhaps antedating neurones, then it is not likely to be subtly affected by nutrients or drugs or anything else. "Learning" disabilities in children are unlikely, therefore, to be the reflection of a defect in the mechanism of learning itself. Does the semantic distinction make a difference? Perhaps. If a child is said to have a learning disability, then there is an implication that the child has a defect in a neurobiological system whose effects may be attenuated by training but which cannot be permanently overcome. If, instead of "learning" disability a neutral term descriptive of the performance deficits is used, the condition seems somehow more tractable. The usual use of the word "learning" in the context of behavioral studies may be misleading, because the results have only remote relations to the basic biological mechanisms of learning, as Crnic commented. Enough of a variety of agents, for example, general anesthetics, will abolish learning, but only along with the abolition of all other behavioral activities. It is not clear that any *selective* effect of any drug on learning has ever been demonstrated.

References

Cravioto J, Delicardie ER (1979) Nutrition, mental development, and learning. In: Falkner F, Tanner JM (eds) Human growth, vol 3, Neurobiology and nutrition. Plenum Press, New York pp 481–511
Dews PB (1957) Studies on behavior. III. Effects of scopolamine on reversal of a discriminatory performance. J Pharmacol Ex Ther 119:343–353
Dews PB (1959) Some observations on an operant in the octopus. J Exp Anal Behav 2:57–63
Dews PB (1986) Dietary pharmacology. Nutr Rev [Suppl] 44:246–251
Dobbing J, Widdowson EM (1965) The effect of undernutrition and subsequent rehabilitation on myelination of rat brain as measured by composition. Brain 88:357–66
Hrboticky N, Leiter LA, Anderson GH (1985) Effect of L-tryptophan on short-term food intake in lean men. Nutr Res 5:595–607
Wurtman RJ, Lieberman HR (eds) (1983) Research strategies for assessing the behavioral effects of foods and nutrients. J Psychiatr Record 17(2):103–230

the wine (unless the effects of flavorants on taste buds is considered pharmac
logy).

It is interesting that none of the chapters in this section directly addre
methods for study of such diet-induced responses, though much of what Rodi
discusses is relevant. Measurement of this type of behavioral response, as b
counting how many customers patronize a restaurant or how many people buy
particular food product, can seem almost trivial in a scientific sense, though no
trivial to the companies whose products are involved. The shallowness of our
understanding of the relevant determining variables is shown by our limited
ability to predict. New food products seem to come to the market by way of a
succession of guesses or intuition, flavor panel testing and test marketing, with
little help from broader scientific generalizations.

Effects of Behavior on Diet

The response to be measured is primarily eating behavior or other behavior
related directly to dietary intake.

Nutritional: an example of a behavioral activity with important effects on diet is
regular, vigorous, prolonged exercise. Lumberjacks typically eat more than
clerks.

Pharmacologic: it is commonplace that behavior-affecting drugs often affect
food intake. This may be due to a direct pharmacologic effect on neural substrates
of eating, as perhaps with amphetamine and related drugs. Often, however, a
drug affects behavior and the altered behavior then leads to changed food intake.
For example, heroin addicts do not eat while under the influence of the drug.

Psychologic effects on diet are well exemplified by the social, temporal and
contextual influences on eating behavior that have been a major field of interest
to Rodin. Such influences can be large. Anorexia on the one hand and most
obesity on the other hand result mainly from such factors.

Conclusions

On Learning

Both Crnic and Thompson refer to "learning." Phenomena of learning have long
been a central theme of experimental psychology. An experiment on the effects
of scopolamine on learning and forgetting performed many years ago engendered
a skepticism about learning as a substrate for drug effects that has persisted,
although the paper has long since been forgotten (Dews 1957). (It was a
progenitor of the type of repeated acquisition paradigm that Thompson advo-
cates.) The following remarks are speculative, perhaps more semantic than
substantive, but the point of view, if accepted, may not be without influence on
the field. Crnic comments: "The ability to learn is so essential to the life of the

Nutritional Aspects of Diet and Behavior Studies of Animals

T. W. Castonguay and J. S. Stern

Introduction

The study of diet and behavior is by its very nature a multidisciplinary one requiring researchers to be competent in a variety of scientific areas including: nutrition, pharmacology, biochemistry, neurochemistry, physiology, psychology, sociology, ethology, ethnology, medicine, computer science, and statistics. Each discipline may emphasize different aspects of the problem and often has its own scientific vocabulary which may not be readily understood by scientists from other disciplines. What is considered "common knowledge" by one group may be ignored by another – a fact which may complicate the experimental designs and interpretation of the data. In the extreme case, it may invalidate the data.

Each discipline has methodologic strengths. The implementation of these methods can result in significant advances in our knowledge of the role of nutrition on behavior. Behavioral scientists, when studying feeding, for example, are careful to control the day/night light cycle. When studying spontaneous activity, they control the access to the experimental room and do not house males and females in the same room. The importance of a dose/response curve in the study of pharmacologic agents is self-evident to the pharmacologist, but has not always been incorporated in some of the early studies of food and its effects on behavior. Appropriate statistical evaluation of data may not be emphasized by those trained in biochemistry and similar disciplines that of themselves often require little in terms of sophisticated analyses of data in order to discover robust effects in the laboratory.

The purpose of this review is to highlight some nutritional considerations required in the formulation of diets for experimental animals used in studies of diet and behavior. These include ways of expressing food or energy intake and the preparation of defined diets. By no means is this review an exhaustive compilation of nutritional factors that can play a role influencing the outcome of studies of diet and behavior. However, we suggest that the factors included are essential for scientists to replicate their observations.

Energy Intake

Many scientists who are concerned with food intake report their data in terms of grams of food eaten. Food intake is often described by the nutritionist in terms of energy intake, that is, gross energy (GE), digestible energy (DE) or metabolizable energy (ME) (see Table A.1 for definition of terms).

Table A.1. Definition of terms

Kilocalorie (kcal): The amount of heat required to raise 1000 grams of water from 16.5°C to 17.5°C. 1 kilocalorie = 4.184 kilojoules

Gross energy (GE): The amount of heat released when a food is completely oxidized in a bomb calorimeter containing 25–35 atm of oxygen

Digestible energy (DE): Gross energy minus fecal energy

Metabolizable energy (ME): Digestible energy minus energy found in urine. ME of a food determined at maintenance corresponds to "physiologic fuel values" found in standard calorie tables. The average value for humans for pure carbohydrate and protein is 4 kcal/g, for fat is 9 kcal/g, and for alcohol is 7 kcal/g

Basal metabolism: The minimum cost of vital processes in a postabsorptive state, under conditions of thermoneutrality and physical and mental rest.

Adapted from Crist et al. (1980).

The following examples are abstracted from the National Academy of Sciences publication *Nutrient Requirements of Laboratory Animals* (1978b). Where appropriate, the original paper is cited. For the rat, digestible energy of a typical low-fat diet is approximately 90%–95% of gross energy; metabolizable energy (ME) is approximately 90%–95% of digestible energy. These figures vary with the type of diet. They are lower when diets are made using natural ingredients (Peterson and Baumgardt 1971a).

Rats will eat a variety of diets and maintain growth and/or body weight. However, if the diets are significantly diluted with inert materials, caloric intake may be insufficient to meet requirements due to the animal's gastrointestinal limitations. The degree of dilution which compromises energy intake varies with the stage of development of the animal. For weanling rats which have a high growth rate, energy requirements are not met with increased food consumption when DE is less than 2.9 kcal DE/cm^3 (Peterson and Baumgardt 1971b). Mature rats, which are growing very slowly have a greater capacity to increase food intake and adapt to lower energy density (less than 2.5 kcal/cm^3). During lactation (a situation where energy needs are greatest), energy density must be at least 4.5 kcal DE/cm^3 (Yang et al. 1969; Peterson and Baumgardt 1971b). As reported by Meyer (1956), non-nutritive cellulose additions to diets decrease DE, in part due to low digestibility of cellulose, but also as a result of decreased digestibility of other dietary components as shown by increased nitrogen losses in the feces.

Finally, no one single growth or reproductive pattern is applicable to all strains of rat. Thus, energy requirements for one strain of rat at any one point in time may be very different from the energy requirement of comparably aged members of the same species derived from a different strain. Evidence from one of our laboratories reveals strain differences in absolute body weight reached by Fischer 344, Osborne–Mendel and Sprague–Dawley rats (McDonald et al. 1988, 1989 and unpublished data).

For studies of energy balance, errors in estimates of energy available to tissues can be made by using the average values for ME for pure protein (4 kcal/g), fat (9 kcal/g), carbohydrate (4 kcal/g), and ethanol (7 kcal/g). Not all carbohydrate sources, for example, yield 4 kcal/g. The average value for carbohydrate is appropriate for cornstarch but is approximately 10% too high for sucrose (3.7 kcal/g) and glucose monohydrate (3.6 kcal/g). Metabolizable energy also varies with species (i.e. ME for alfalfa meal for swine is 1.7 kcal/g and for cattle is 2.2 kcal/g) (National Academy of Sciences 1979, 1989).

While few readers of this review are likely to be studying behavior in swine and cattle, many may study food intake in diabetic animals. When fed a diet high in carbohydrate, diabetic animals are hyperphagic, that is, they eat more grams of food than their nondiabetic controls. However ME intake (which takes into account the variable amount of glucose lost in the urine) may be comparable (Chan et al. 1982).

Defined Diets and Replication of Experiments

One of the most fundamental aspects of contemporary science is the demand for the experimenter to define specific conditions in which a cause – effect observation is made. If, upon repeating the conditions of the experiment, the same relationships are not observed, the reliability of the first observations is deemed suspect and may be invalid. Sidman (1960) has argued that defining conditions in which observations are made is an important part of the behavioral experiment. Furthermore, failure to replicate experimental conditions may mean that the experimenter has failed to account for all the pertinent variables affecting the outcome of the experimental trial. It is assumed that if all of the important variables are controlled, the outcome of the second and subsequent trials will match the initial observation.

In general, defining stimuli as carefully as possible has been one of the strengths of modern psychophysics and experimental psychology. Yet, this practice has not always been followed by the same scientists when studying nutrition and behavior. Over the past decade, several lines of research have been propounded that have made use of dietary regimens which vary widely in composition and caloric density. It has been argued that the use of imprecisely defined diets inevitably increases the variability of observed responses, thereby potentially masking more subtle effects, and increasing the likelihood of subsequent replications yielding differing results (Castonguay 1987; Moore 1987). This argument applies to laboratory practices as well as to clinical and field studies.

Types of diets for laboratory animals include: "natural ingredient" diets which are made from ingredients such as whole grains (e.g. stock diets such as Purina rat chow; see Table A.2); "purified" diets which are made from refined ingredients such as casein (protein source), starch (carbohydrate source), corn oil (fat source); and "chemically defined" diets made from amino acids, sugars, essential fatty acids, etc. (National Academy of Sciences 1978b). In addition, "cafeteria" or "supermarket" diets (containing commercial products such as cookies and luncheon meats) have also been used (Sclafani 1976).

Table A.2. Printed information found on each 50 lb bag of Purina Rat Chow No. 5001

Guaranteed analysis
Crude protein not less than 23.0%
Crude fat not less than 4.5%
Crude fiber not more than 6.0%
Ash not more than 8.0%
Added minerals not more than 2.5%

Ingredient label
Ground extruded corn, soybean meal, dried beet pulp, fish meal, ground oats, brewers dried yeast, alfalfa meal, cane molasses, wheatgerm meal, dried whey, meat meal, animal fat preserved with BHA, dicalcium phosphate, salt, wheat middlings, calcium carbonate, vitamin B_{12} supplement, DL-methionine, calcium pantothenate, choline chloride, folic acid, riboflavin supplement, thiamin, niacin supplement, pyridoxine hydrochloride, ferrous sulfate, vitamin A supplement, D-activated animal sterol, vitamin E supplement, calcium iodate, ferrous carbonate, manganous oxide, cobalt carbonate, copper sulfate, zinc sulfate, zinc oxide

Many investigators prefer to purchase diets used in experimental animal studies, primarily for reasons of convenience if diet mixing facilities are not available. Unfortunately, commercial diets (i.e., those that can be purchased from feed or chemical suppliers) have several disadvantages. The ingredients of stock diets, or chows are formulated to meet the recommended nutrient intakes of the animal, but are usually composed of ingredient sources which may change from batch to batch. Diets of more defined composition are also commercially available, but may not meet either the requirements of the animal or the current experimental needs of the investigator. This is especially true if the diets were first formulated several years ago. No diet should be purchased unless its composition is carefully compared with the requirements of the animal being studied. Finally, a major disadvantage of purchased diets is that the investigator often has only a limited opportunity to manipulate the content of the nutrient under study.

The use of nutritionally adequate diets is obviously important when studying diet and behavior. Nutritional requirements vary with the animal species, strain, and sex, and are influenced by physiologic states such as growth and reproduction. The nutritional requirements of an animal reflect thousands of years of natural selection pressures that were and are present in the ecological niche of each species. Hence the dietary requirements of the mouse differ from those of the rat, despite the fact that both are rodents. Similarly, goats, sheep, and cattle, which are all ruminants, have nutrient needs that differ from one another. Not only are there differences in the requirements among species, but different strains of the same species can have different responses to some nutrients. For example, an Osborne–Mendel rat can become obese when fed a high fat diet, whereas the same diet has no effect on weight gain or adiposity of the Fischer 344 rat. To make matters yet more complicated, within each animal is a shifting baseline of nutrient requirements, so that levels of some nutrients (e.g., protein) can be high early in the animal's lifetime relative to its need for that same nutrient later in life. Similarly, the sex of the animal can influence nutrient need (see Table A.3). The National Academy of Sciences publishes, and continues to update a series of booklets on the nutrient requirements of a variety of animals (National Academy of Sciences 1978a,b, 1979, 1985a,b, 1989).

Perhaps the most popular example of the lack of dietary definition or standards has been the use of the "cafeteria" feeding regimen in the promotion of diet-

Table A.3. Nutritional requirements are influenced by a number of factors including species, strain, age, sex

---

Species
Simple systems
Small animals: rat, mouse, rabbit, guinea pig, hamster, cat, dog
Ruminants: sheep, goats, cattle
Primates – non-human primates, humans

Strain and substrain
Rat: Fischer 344, Long Evans, Sprague Dawley, Osborne Mendel, genetically obese (*fa/fa*) and lean (*Fa/Fa*) Zucker rats

Age
Weanling, young, adolescent, adult, aged

Sex
Male, female (pre-, post-menopausal)

Physiological state
Pregnancy, lactation, exercised, sedentary

---

induced overeating and obesity. Sclafani and his associates were among the first to re-examine the precision with which laboratory rats control their daily caloric intake (Sclafani 1976). Unlike the notable day-to-day precision with which their animals ate composite laboratory diet in balance with their energy and nutrient requirements, rats given access to a variety of food items, ranging from salami to chocolate chip cookies, failed to preserve caloric intake. Rats significantly increased their daily caloric intake, and became obese. These studies were seminal in stimulating researchers to explore the mechanisms that control food intake. However, the methodology involved does not support more rigorous analysis by virtue of the experimental error induced by trying to estimate composition of intake. For example, the caloric and nutrient composition of salami may vary with its manufacturer. Similarly, there are significant differences in the compositional aspects of chocolate chip cookies from different sources, as with any baked goods. These variants all contribute to the introduction of experimental error in the quantification of intake that can only be eliminated by careful composition studies of samples from each of the foods offered during each experiment. This significant requirement is both expensive and time consuming.

Some researchers have continued to use "cafeteria diets" when studying phenomena such as non-shivering thermogenesis and oxygen consumption and other aspects of metabolism. The pursuit of mechanistic hypotheses under these variable or less well-defined dietary conditions is doomed to controversy, if for no other reason than the fact that other investigators, using food items that also vary in composition, may fail to replicate all but the most robust effects.

Nonetheless, the use of definable diets has its drawbacks. Defined diets are relatively expensive, they are more difficult to prepare, and they may not be as palatable as commercially prepared foods. There is also the question of whether or not formulated semi-synthetic diets are truly physiologic, since they may require less digestion, and their absorption may be different from more natural diets. On the other hand, experiments are even more expensive to replicate, especially if they fail to confirm the initial findings. In these times of restricted funding for research and increased scrutiny from the public about the conduct of "needless replication" in animal studies, long-term cost-effective strategies are supportive of the use of known, definable diets in place of less expensive, but compositionally unknown or undefined food sources.

Natural ingredient or stock diets are routinely used in place of nutritionally defined diets for routine maintenance of experimental animals. While there is no question that manufacturers of stock laboratory diets typically do an excellent job in the formulation of their products, it nevertheless remains an active concern when these diets are used in experimental protocols. The reason for this concern stems from the compositional changes that are repeatedly made by manufacturers so as to guarantee minimum basic standards and at the same time optimize profits from the sale of the product when the cost of agricultural products varies dramatically over time. Thus, stock diets can vary in composition, while at the same time maintaining comparable appearance, and being marketed under the same name (Table A.2). Not only are there compositional differences between bags of stock diet (i.e. different batches), but the age of the stock diet and the storage conditions can vary, ultimately leading to significant decreases in vitamin content and protein quality. For example, a bag of Purina rat chow manufactured in St. Louis, Missouri, in the middle of winter may be transported to East Lansing, Michigan, in a relatively short time, with minimal losses from sitting in a box car. However, another bag of chow, with the same nutrient content at the factory, may be shipped in midsummer to Davis, California, and may be subject to much more heat, causing some protein denaturation and vitamin oxidation. Thus, while both researchers use the same brand name stock diet, their animals will be provided with different nutrient levels. The potential differences in experimental results could be eliminated by the use of diets made from assayed sources, put together for use at the time of the experiment, and stored under appropriate conditions (Fullerton et al. 1982).

The inclusion of diet composition as one of the environmental variables in nutrition and behavioral research, should be consistent with the concept of scientific rigor for behavioral studies proposed by Sidman (1960).

Other Dietary Considerations

Emphasis on the use of standardized, uniform diets in the conduct of experiments focusing on nutrition and behavior is an extension of similar applications of the principle to other pursuits, such as the study of metabolism or endocrine function. There are other advantages associated with defined diets that are not as obvious. One such advantage is that the use of formulated defined diets allows the experimenter to deliberately vary the composition of one component of the diet, while maintaining other aspects of the diet. For example, the concentration of protein can be systematically varied without introducing concomitant changes in caloric density, mineral or vitamin concentrations. This is especially important for studies of rapidly growing animals and pregnant animals. Similarly, variations in the caloric density of the diet can be achieved without sacrificing the balance between macronutrients.

There is no better example of how systematic variation in diet composition can be used in nutrition and behavior experiments than the way in which diet affects central nervous system function. It has been known for many years that variations in the concentration of protein can affect neurotransmitter concentrations which, in turn, presumably potentially alter behavior; though we should also note that the relationship between changes in neurotransmitter concentration and particular behavioral states is not defined (Peters and Harper 1987). For example, there are several reports that increases in circulating tryptophan can induce increases in brain serotonin (Fernstrom and Wurtman 1971). Increased serotonergic activity has been associated with decreased latency to sleep, leading to the clinical suggestion that drowsiness can be brought on by consumption of small amounts of tryptophan shortly before retiring for the evening (Hartmann and Greenwald 1984). Although this suggestion has been made and put into clinical practice, it remains to be definitively established that the specific increase in tryptophan that led to an increase in serotonin caused increased drowsiness by virtue of increased serotonergic transmission. The careful design and implementation of dietary supplementation experiments can only be conducted under specifically defined conditions.

In addition to considerations of protein and calorie content of experimental diets, all components of the diets should be as well specified as possible. For example, fat sources should be listed separately and, when possible, the fatty acid composition of these sources should be stated. Simlarly, the use of dietary fiber sources should be clearly detailed. Standard concentrations for these and the other essential macro – and micronutrients should be clearly detailed.

Not only are the species-specific requirements important to include in the design of the experimental diet, but it also becomes important to have a working knowledge of the dynamics of the individual food items or constituents of the diet. For example, it becomes very important to correct the tocopherol concentration of a diet that has been supplemented by the addition of some types of fat. Fat sources that are high in polyunsaturated fatty acids require additional vitamin E to retard oxidation and subsequent rancidity. Diets high in polyunsaturated fatty acids also increase the animal's requirement for vitamin E.

The consideration of the different nutritional needs of one's experimental animals introduces another set of factors that play major roles in the outcome of any nutrition/behavior experiment. For example, rats eat predominantly at night. Several investigators have reported that the typical laboratory rat given ad libitum access to food will eat up to 85% of their total daily caloric intake during the dark half of a 12/12 hour day/night cycle. Thus, conducting a test using food as a reward may be biasing its outcome if the test takes place at the "wrong" time of day. Food rewards may be much more effective if the animal is ready to eat. Conversely, food may not serve as a good reward if the animal has no tendency to eat. Depriving a rat of food for 12 hours during the day or light cycle is not a comparable degree of deprivation to depriving the same rat during the 12 hours of the night or dark cycle, when feeding normally occurs.

Similarly, what is reinforcing for one species may be just the opposite for another. For example, dogs prefer a moderate degree of rancid fat in their diets. That same level of rancidity in the diet is avoided by cats (Castonguay, unpublished observations). Salty solutions given to an adrenalectomized rat are consumed with avidity, whereas salty solutions given to adrenalectomized hamsters are met with indifference (T. Bartness, personal communication). These and other examples illustrate the point that an appreciation of the species under study should be an important cornerstone of this type of research.

Finally, procedural elements of this type of research may dramatically affect outcomes. For example, the routine maintenance conducted in these experiments usually includes the determination of amount of food eaten. Spillage is collected and weighed, and each animal typically has food that remains from the previous maintenance period. If that diet has been marked with urine it can alter collection figures. If the diet is composed of enough fat that it has started to go rancid, replacing the diet with fresh diet, as opposed to simply adding to the food cup, will stimulate eating.

Conclusions

This chapter briefly reviews some of the factors that nutritionists consider when formulating diets for experimental animals. For more details, the reader should consult publications issued by the National Academy of Sciences which discuss, in depth, the nutritional needs of individual species.

Acknowledgements. The authors would like to thank Drs. Eleanor Williams, Phylis Moser-Veillon, Richard Ahrens and Ms. Marie DeStefano for their critical reading of the manuscript. This chapter was supported in part by grants DK-18899, T32 DK-07355, DK-35747 and the Whitehall Foundation.

References

Castonguay TW (1987) Diet selection: principles, rules, and suggestions. In: Toates F, Rowland N (eds) Feeding and drinking: techniques in the behavioral and neural sciences, vol 1. Elsevier, Basel, pp 429–442

Chan CP, Koong LJ, Stern JS (1982) Effect of insulin on fat and protein deposition in diabetic lean and obese rats. Am J Physiol 242:E19–E24

Crist K, Baldwin RL, Stern JS (1980) Energetics and the demands for maintenance. In: Alfin-Slater R, Kritchevsky (eds) Nutrition and the adult. Plenum, New York, pp 159–182 (Macronutrients 3)

Fernstrom JD, Wurtman RJ (1971) Brain serotonin content: physiological dependence on plasma tryptophan levels. Science 173:149–152

Fullerton FR, Greenman DL, Kendall DC (1982) Effects of storage conditions on nutritional qualities of semipurified (AIN-76) and natural ingredient (NIH-07) diets. J Nutr 112:567–573

Hartmann E, Greenwald D (1984) Tryptophan and human sleep: An analysis of 43 studies. In: Schlossberger HG, Kochen W, Linzen B, Steinhart H (eds) Progress in tryptophan and serotonin research. Walter de Gruyter, Berlin, pp 297–304

McDonald RB, Horwitz B, Stern JS (1988) Cold-induced thermogenesis in younger and older Fischer 344 rats following exercise training. Am J Physiol 254:R908–R916

McDonald RB, Stern JS, Horwitz BA (1989) Thermogenic younger and older rats to cold exposure: comparison of two strains. J Gerontol 44:B37–B42

Meyer JH (1956) Influence of dietary fiber on metabolic and endogenous nitrogen excretion. J Nutr 58:407–413

Moore BJ (1987) The cafeteria diet – An inappropriate tool for studies of thermogenesis. J Nutr 117:227–231

National Academy of Sciences (1978a) Nutrient requirements of cats, rev edn. National Academy Press, Washington, DC

National Academy of Sciences (1978b) Nutrient requirements of laboratory animals, 3rd rev edn. National Academy Press, Washington, DC

National Academy of Sciences (1979) Nutrient requirements of swine, 8th rev edn. National Academy Press, Washington, DC
National Academy of Sciences (1985a) Nutrient requirements of dogs. National Academy Press, Washington, DC
National Academy of Sciences (1985b) Nutrient requirements of sheep, 6th rev edn. National Academy Press, Washington, DC
National Academy of Sciences (1989) Nutrient requirements of dairy cattle, 6th rev edn. National Academy Press, Washington, DC
Peters JC, Harper AE (1987) A sceptical view of the role of central serotonin in the selection and intake of protein. Appetite 8:206–210
Peterson AD, Baumgardt BR (1971a) Food and energy intake of rats fed diets varying in energy concentration and density. J Nutr 101:1057–1068
Peterson AD, Baumgardt BR (1971b) Influence of level of energy demand on the ability of rats to compensate for diet dilution. J Nutr 101:1069–1074
Sclafani A (1976) Appetite and hunger in experimental obesity syndromes. In: Novin D, Wyrwicka W, Bray G (eds) Hunger: basic mechanisms and clinical implications. Plenum Press, New York, pp 281–296
Sidman M (1960) Tactics of scientific research. Basic Books, New York
Yang MG, Manoharan K, Young AK (1969) Influence and degradation of dietary cellulose in cecum of rats. J Nutr 97:260–264

Section III

Epidemiologic Studies

Introduction

A. P. Simopoulos

In a discussion of methodologies it is essential to consider the epidemiologic methods used in assessing relationships between diet and behavior. The evidence regarding health hazards related to environmental and lifestyle factors generally comes from epidemiologic studies based on nonexperimental data. Therefore it should be said from the outset that epidemiology as a science has not advanced to the level of the biological sciences that include the study of animals or molecules and the methods used in epidemiology have not yet reached a level of scientific excellence.

Feinstein (1987) in his paper on "Scientific standards and epidemiologic methods" has identified and extensively discussed the difficulties and problems involved in epidemiologic studies. These involve the structure of experimental studies; the selection of groups of data; and the quality of basic data. Bias in epidemiologic studies is a major problem that ought to be taken into consideration. There are problems with susceptibility bias; problems with ascertainment bias, and bias in performance of maneuvers. There is therefore a need to establish standards of vigilance and other research activities that can improve the state of this science in the field.

Furthermore, genetic variability in the population ought to be taken into consideration. The genetic aspect needs to be brought to the attention of epidemiologists. Family factors ought to be considered, and predisposition or susceptibility to disease ought to be part of the armamentarium of epidemiologic research. Because of genetic variation, individuals in the same environment or exposed to the same environmental factors respond differently. Genetic epidemiology should improve our understanding of the genetic contribution to the interaction between diet and behavior and eventually distinguish between cause and effect.

In carrying out studies on nutrition and behavior we should consider the fact that the absorption of nutrients depends on a number of processes that are genetically controlled and that significant population differences for many gene-determined common enzyme deficiencies exist. Therefore, studies carried out in one ethnic group should not be extrapolated to another ethnic group unless there is evidence that the ethnic groups do not differ significantly in their genetic profile.

In addition to evaluating the state of the art of epidemiologic methods in diet and behavior, and defining both the methods used in nutritional epidemiology and the methods used in measuring behavioral outcomes, it is essential that this new field of epidemiology, namely genetic epidemiology and its contribution to diet and behavior studies, is also considered.

In their chapter, Rhoads and colleagues discuss the different types of studies that show an association between diets and the development of specific health problems. Three general types are considered: (1) studies which show correlations between disease frequency and diet among different communities or nations; (2) studies involving the individual's diet and the presence or development of disease; and (3) experimental studies in which diets or dietary constituents are fed to volunteers and their effect on health or behavior is determined. Of the three types, experimental studies are the most scientific: therefore, in studies involving diet and behavior, experimental approaches should be used whenever possible.

In their chapter on "Principles of behavioral epidemiology", Vietze and Kiely discuss the methods used in obtaining psychologically salient data in large-sample epidemiologic studies and the need to develop valid methods. Concerns are further illustrated by discussing epidemiologic problems from the fields of developmental and behavioral psychology; a number of examples are provided.

The chapter "Identifying the genetic component of dietary behavioral interactions: a challenge for genetic epidemiology" by Ward, presents an extensive review and critique of the methods that have been used in defining genetics in lipid abnormalities, obesity, alcoholism and lactose intolerance. Ward points out that three important issues need to be resolved for the successful identification of the role played by genetic factors. These are: An appropriate model needs to be defined; the variables to be measured need to be identified and they should be relevant and objective; and an effective analytical strategy needs to be utilized. Dr Ward presents a paradigm for the influence of genes and environment on dietary behavioral interactions that takes metabolic variability as the fulcrum through which genetic factors mediate this interaction. Using the principles of genetic epidemiology, it should be possible to improve understanding of the genetic contribution to the dietary–behavioral interactions and eventually define cause and effect relationships. Ward comments that twin studies, path analysis and the application of regressive techniques to pedigree structured data offer considerable promise, especially if potentially confounding environmental variables are measured. With the advent of a well-defined genetic map, linkage analysis has great potential, but requires that the correct genomic segment be selected for study.

In the discussion of this section Harper expands on the principles and issues identified by the contributors and further emphasizes the need for genetic concepts and genetic epidemiology to be considered essential to our understanding of the relationship between diet and behavior.

Acknowledgement. This work was supported in part by the Howard Heinz Endowment.

Reference

Feinstein AR (1987) Scientific standards and epidemiologic methods. Am J Clin Nutr 45 [suppl]:1080–1088

Chapter 7

Nutritional Influences on Childhood Behavior: the Epidemiologic Approach

G. G. Rhoads, F. A. Rhoads and P. H. Shiono

Epidemiologic observation has contributed substantially to our understanding of the relation of diet to disease. The classic studies of pellagra illustrate the power of careful observation of groups of people in developing causal hypotheses, as well as the power of the laboratory in elucidating actual mechanisms of disease pathogenesis. A variety of study approaches was used to unravel the role of niacin deficiency in pellagra, but the most persuasive was the simple observation (simple after the correct hypothesis was formulated) that supplementation of the diet of affected individuals with niacin-containing food reversed the illness. While it is not as easy to conduct such powerful and convincing demonstrations of the role of diet in diseases associated with a surfeit of nutrients (or of food additives), nevertheless epidemiologic observation is a widely used tool. This chapter will briefly review the main approaches which have been used in nutritional epidemiologic studies and will explore their promise for testing hypotheses about nutrition and behavior in children.

Observational Methods

Epidemic disease may at times be a striking phenomenon, demanding study in urgent circumstances. It is hardly surprising, therefore, that some epidemiologic methods which have come into use for these purposes are "rough and ready." However, as the focus of epidemiologic investigations has shifted to chronic diseases and other health conditions, and as the size of risks being sought has diminished, study designs have become progressively more sophisticated and now include randomized clinical trials which can be quite similar to randomized designs used in experimental psychology. Epidemiologic studies can be conveniently divided into three general types (Table 7.1).

Descriptive and ecologic studies comprise the simplest of these categories and are used mainly to formulate causal hypotheses, or to make predictions about the

Table 7.1. Epidemiologic study designs

Descriptive and ecologic studies

Observational studies of individuals
    Cohort (prospective)
    Cross-sectional
    Case–control (retrospective)

Clinical trials (challenge tests)

---

future frequency of disease. Descriptive studies typically examine affected individuals with respect to a number of key variables including their personal characteristics (age, sex, race, socioeconomic status), the timing of the occurrence of their illness, and the place where it occurs or may have been acquired. A typical example would be the investigation of a food poisoning outbreak, where the characteristics of person, place and time form the basis for inferring a cause of the resulting sickness.

Ecologic studies extend the concept of descriptive studies by looking at the characteristics of *groups* of persons or populations that are disposed to a certain disease rather than looking at specific affected individuals. The prevalence of suspected dietary or other risk factors among the groups is correlated with the frequency of the health outcome of interest. We know, for instance, that dental caries is common in societies where sugar consumption is high and that heart disease occurs more frequently in populations with high animal fat intakes. The difficulty with these observed correlations is that they may often be spurious. Thus, on an international basis, sugar is associated with coronary disease nearly as well as is fat (Yudkin 1964), but there is little supporting evidence that it causes heart attacks (Bierman 1979). For this reason ecologic studies, though widely used to generate hypotheses, rarely, if ever, provide a critical test of their validity.

A second general category of epidemiologic study is the family of analytic observational designs, in which individuals are studied to see whether particular characteristics or risk factors are associated with a disease of interest. This type of design has been used extensively to try to link dietary factors with chronic disease. These studies can be conveniently classified into three main types: prospective or cohort studies, in which the diet is characterized before the health problem develops; cross-sectional studies, in which the diet and the health condition are ascertained simultaneously; and case–control or retrospective studies in which a number of individuals with the disease and a series of comparable controls without the disease are compared in terms of their diets. Conceptually, the simplest and most robust of these designs is the cohort study, where diet is assessed before the disease or health condition develops. Since there is little or no opportunity for the disease to affect dietary intake, this has the advantage that the direction of any causal relationship is clear. In the cross-sectional study, where disease and diet are ascertained simultaneously, there is always the possibility that diet may be affected by the presence of disease. This would certainly be the case for studies of diet and hyperactivity in children, where an active child might eat more, or might be more inclined than a passive child to snack in a more uncontrolled way.

Case–control studies, the third major observational category, differ from the others by focusing on persons with the disease of interest and comparing them with a sample of non-diseased persons. This design is particularly useful when the health condition is rare, making prospective or cross-sectional study of a general population impractical. Case–control studies have been extensively used in examining possible dietary precursors of less common types of cancer, which would occur only rarely in a prospective study of reasonable size. The main difficulty with retrospective designs is that in order to focus on the disease of interest, the condition must already have occurred; that is, one must compare persons who already have the disease with a presumably comparable group of persons who do not. The opportunity for reverse causality, where the presence of the disease affects the diet, its recall, or its measurement, is obvious. Moreover, all of these observational study designs are subject to confounding of the diet–disease relationships by other factors such as age, sex, socioeconomic status and other variables which may be related to diet and also affect the frequency of disease.

Observational studies linking diet to chronic disease among individuals within a single population have, in general, been disappointing. In the cardiovascular area, for instance, prospective studies such as that at Framingham were successful in identifying the importance of smoking, of low density lipoprotein (LDL) cholesterol levels, of high density lipoprotein (HDL) cholesterol levels, and of blood pressure in the genesis of coronary heart disease. But when the diets of persons developing coronary disease were compared with those of persons who remained well, there was almost no difference (Gordon 1970). This does not mean that the consumption of saturated fat is not important in the genesis of the disease. We know that consumption of these fatty acids causes an increase in LDL cholesterol, and we know that LDL cholesterol is linked to atherosclerosis. The failure, rather, appears to be one of dietary measurement.

For a measurement to characterize an individual's risk of disease, it must be able to separate him or her from others in the same population. In statistical terms, this means that the variation in the measurement *between* individuals must be large relative to the variation in the measurement *within* individuals. This ratio of variances is reflected in the correlation coefficient which can be calculated between repeated measures. Homogeneity of diet within a population will diminish the between-individual variance and will tend to lower this correlation. Poor measurement has a similar effect. Table 7.2 shows these correlation coefficients for several cardiovascular risk factors which were assessed 2 years apart at the Honolulu Heart Study (Rhoads 1987). The blood pressure and serum cholesterol measures at the baseline exam were highly correlated with repeat assessments done two years later, values being about 0.7. Uric acid, which is sometimes found to be a risk factor and sometimes not, had a correlation of about 0.5.

Table 7.2. Reproducibility of selected non-dietary risk factors measured 2 years apart (Honolulu Heart Study)

Risk factor	Number of observations	Correlation coefficient
Systolic blood pressure	7500	0.70
Diastolic blood pressure	7500	0.67
Serum cholesterol	1301	0.71
Uric acid	1301	0.53

Table 7.3. Correlation of 24-hour recall nutrient estimates with repeat estimates 2 years later in 318 Honolulu Heart Study participants

Nutrient	Method of repeat estimate	
Calories (kcal)	0.38	0.49
% Calories as fat	0.19	0.45
%Calories as PFA	0.05	0.16
%Calories as alcohol	0.63	0.74
Starch (g)	0.55	0.64
Cholesterol (mg)	0.27	0.35

Table 7.3 shows similar correlations for a series of nutrients which were measured by 24-hour recall at the first examination and validated against a 7 day diary and repeat 24-hour recall after a 2 year interval. The main hypotheses in this study of diet and cardiovascular disease were related to fat and cholesterol; but it is clear that participants were not adequately categorized in terms of these variables. It is not surprising then, that little relation of these variables to coronary heart disease could be demonstrated (Yano et al. 1978). Alcohol, which was better measured, had quite a striking (inverse) relation to coronary disease in this study (Yano et al. 1977). The use of food intake frequency data provides somewhat better correlations for repeated measures than does the 24-hour recall but these are still not as good as the reproducibility of the standard risk factors (Rhoads 1987). Of course, a strong correlation between repeated measures does not guarantee that a measurement is valid. Two measures can be in excellent agreement but both be wrong. However, when a measure does not yield good reproducibility it cannot be a valid assessment of average intake.

What does all this have to do with diet and behavior? It has been shown that observational studies have been rather unsuccessful in relating diet to chronic diseases, at least in part because of substantial difficulties in classifying people by what they eat. This is partly a problem of measurement but is also partly because of the relative homogeneity of diet within particular populations. These same problems will affect any attempt to relate usual diet to chronic behavioral conditions or mental illness. When possible, such studies should be based on biochemical measures of nutritional status rather than the usual food frequency and diary methods.

Experimental Studies

Perhaps the main conclusion to be drawn from reviewing these dietary studies of other diseases is that one should try to avoid the observational approach in favor of clinical trials in which dietary intake is directly manipulated to examine its effect on specific behavioral outcomes. Whereas this is not feasible for cancer and heart disease because of the very long latency periods of these conditions, many of the diet–behavior hypotheses involve short-term effects which can be encompassed in practical experiments. Such experiments should incorporate as many of the usual design safeguards as is feasible. Subjects should be randomly allocated to treatment groups to avoid confounding by extraneous variables.

Participants and staff should be kept unaware of group assignments (masked assignment) to ensure that their behavior and reporting are not influenced by subjective biases. Masked assessment assures that those who are observing or measuring the outcome behavior cannot introduce a conscious or unconscious bias for (or against) the hypothesis. These precautions are desirable in any randomized trial, but they are especially important where the participant's behavior could be influenced in an important way by knowing the assignment or where endpoints have a potentially subjective component.

Some clinical trials make use of crossover designs in which participants are randomly assigned to time periods with active intervention and with placebo, thus serving as their own control. These designs are statistically powerful and improve the chance of getting a significant result with a limited number of subjects. Their interpretation can occasionally be complicated when different results are obtained depending on the order in which placebo and active drug are introduced.

An Example: Food Additives and Childhood Hyperactivity

A controversial area in diet and behavior, which has been the subject of several small, well-designed trials as well as much provocative speculation and anecdote is Feingold's hypothesis that food colors and salicylates cause and/or exacerbate attention deficit hyperactivity disorders (ADHD) in young children. Although the controlled studies have shown little clear evidence of benefit from the Feingold elimination diet, the dietary therapy for such children remains widespread. Varley reported that among 100 families attending a clinic for children with ADHD, 80 had attempted to implement a diet, and more than a third of them felt that it had been helpful (Varley 1984).

One of the early well-designed studies of this issue was undertaken by Harley and colleagues (1978a). They tested an elimination diet for 46 boys with ADHD using a crossover design in which half the children were put first on the experimental diet and half on placebo. All medication was stopped. After 3–4 weeks the diets were switched. The following description of their methods provides a sense of the effort that is required to implement such a study.

Several steps were taken to maximize compliance [with the diet], obviously a critical factor. First all the investigators met with the participating families in a general meeting, and while the importance of compliance was stressed, the families were informed that certain infractions of the diet would undoubtedly occur, and such infractions should be carefully documented and reported. Second, the dietitians made initial individual home visits to ascertain family eating habits, reinforce the importance of compliance, and instruct the parents in maintaining dietary records. Finally, arrangements were made to have all previously purchased foods removed from the house and to have each family's entire food supplies delivered to their homes weekly.

All family members were placed on the diet to minimize the possible treatment effects related to the experimental subject and his special diet, and also to reduce his temptation to eat other foods that would ordinarily be available to "non-involved" family members. The weekly food deliveries also contributed to the blind aspect of the project, since families were informed that they would be on various diets over a six- to eight-week period and were not told they would be on one of two diets. In addition to providing the family's ordinary food needs, supplementary food was delivered for special occasions such as holidays, guests, family dinners, etc. At school, children customarily provided treats or snacks for their classmates on their birthdays. To avoid diet disruption in these situations we made arrangements to deliver approved treats to the entire class when any child in the room had a birthday.

Several other procedures were included to obscure the diet manipulations. Special production runs were made to prepare identically packaged chocolate bars and specialty cakes, with one containing standard ingredients and the other free of artificial flavors or colors. The production and coding of these specially prepared food items were directly supervised by one of the authors. Also, a number of pseudo-dietary manipulations and distractions were incorporated into the diets. For example, the family might be provided with hot dogs, potato chips, and cookies one week, and these items would be absent from the next week's menu. This might be interpreted to the child and/or his parents as evidence of being on two distinct diets, but these items (depending on brand selection) were permitted on both the control and experimental diets. Finally, sweet potatoes were systematically introduced and removed throughout the dietary phase as another pseudo-manipulation distracting technique.

Changes in the children's behavior were assessed by extensive, structured classroom observation and by a number of psychologic tests including the Wechsler Intelligence Scale for Children and the Wide Range Achievement Test. Compliance with the dietary intervention was assessed through weekly visits to the homes by study nutritionists who reviewed dietary records and by reports from parents and teachers about infractions, which averaged about one per week per child.

Although the authors found no evidence on teacher rating scales or psychologic testing of an overall benefit from the diet, they reasoned that this might be obscured if only a minority of hyperactive children were adversely affected by food additives. Accordingly, they selected the 9 boys whose scores had improved the most on the diet and continued the elimination protocol for an additional 9 weeks (Harley et al. 1978b). During this time food dye challenges were alternately added and deleted for 2–3 week periods in a masked fashion while observation of the subjects was continued. A 27 mg challenge dose of mixed common food dyes was used and was believed to represent a common level of exposure among American children. No deleterious behavioral effects of these challenges were observed on the boys' behavior.

In another well-planned investigation Weiss and co-workers studied 22 children aged 2–7 years with ADHD who were alleged to be responsive to an elimination diet (Weiss et al. 1980). After instituting a diet that excluded artificial colors and flavors they arranged for each child to consume a bottle of soft drink at a specified time each day for 77 days. The drink usually contained only caramel and cranberry coloring (placebo), but on 8 days distributed randomly among weeks 3 through 10 a challenge drink, indistinguishable by sight, smell, taste, or color, but containing 37 mg of food dyes, was substituted without the knowledge of the children, the parents, or the study staff. Each day parents scored selected aversive behavior for two 15-minute observation periods, one within 3.5 hours after the soft drink consumption and one later in the day. In addition, global estimates of the frequency of these behaviors and the ten-item Connors Parent Rating Scale were completed daily.

In 20 of the 22 children there was no evidence of behavioral change after the challenges. One 3-year-old boy displayed significant elevations in two target behaviors which were previously emphasized by the mother as most characteristic of dietary infractions. A 34-month-old girl was reported to react dramatically; highly significant increases were noted in most of the target behaviors and in all global assessments. Moreover on 6 days the mother spontaneously commented that she thought the child had been challenged. She was correct on 5 occasions ($P = 1.6 \times 10^{-5}$). Apparently no attempt was made to ascertain which of seven dyes included in the challenge drink was responsible for this behavioral effect.

Swanson and Kinsbourne studied 40 children with "behavioral symptoms suggesting hyperactivity", including 20 who had shown a favourable response to stimulant medication and 20 "in whom an adverse response had been documented" (Swanson and Kinsbourne 1980). Based on the Connors Parent Rating Scale and other information the first group was felt to have confirmed hyperactivity while the second group was not. The children were admitted in pairs to a clinical investigation unit for controlled implementation of the Feingold diet. All medication was stopped the day before admission and the diet was strictly maintained for 3 days. At 10.00 a.m. on days 4 and 5 the children received capsules that contained either 100–150 mg of food dyes or placebo, the order being counterbalanced across subjects. Paired-associate learning tests were administered at 9.30, 10.30 and 11.30 a.m. and at 1.30 p.m. and revealed an increased number of errors 1.5 to 3.5 hours after this large dose of food dye (but not after the placebo), only in the 20 children who were drug responsive. There was no effect in the other children and no information was provided about the effect of ordering of challenges and placebo on the results.

These three studies illustrate some of the approaches to sophisticated double masked challenge tests that can be used to untangle the short-term behavioral effects of food dyes. While the studies suggest that modest amounts of food dye do not cause or exacerbate attention deficit hyperactivity disorder in most affected children, they may do so in a small minority. The possibility that larger doses can interfere with certain types of learning (Swanson and Kinsbourne 1980) remains unconfirmed. These studies illustrate that it is technically feasible to study short-term behavioral effects of food additives but additional studies with larger sample sizes are needed to clarify the role of specific substances. As illustrated by the detailed descriptions of methods given above, these studies are not easy to do well. Nor will they ever completely lay to rest the suspicions of parents or doctors who believe a particular child has responded to an elimination diet. Nevertheless, in view of the widespread use of food additives more replication of this work and more attention to the effects of specific ingredients would be desirable.

Conclusion

We believe that the role of the traditional observational epidemiologic study in this area is quite limited. Where long-term effects are suspected, making randomized clinical trials impossible, biochemical markers of nutritional status are likely to produce results that are more valid and more easily reproduced than dietary recall assessments. The latter have proven difficult in the context of other chronic diseases and uncertainty in documenting a specific time of onset of behavioral disorders adds further challenges to the utility of these approaches. Doubtless there will be instances where long-term observational studies of the effect of diet on behavior will seem necessary. But let them be a last resort.

References

Bierman EL (1979) Carbohydrates, sucrose, and human disease. Am J Clin Nutr 32:2712–2722
Gordon T (1970) The Framingham diet study. In: Kannel WE, Gordon T (eds) The Framingham study. An epidemiological study of cardiovascular disease, section 24. Public Health Service, DHHS, Washington. US Government Printing Office

Harley JP, Ray RS, Tomasi L et al. (1978a) Hyperkinesis and food additives: testing the Feingold hypothesis. Pediatrics 61:818–828
Harley JP, Matthews CG, Eichman P (1978b) Synthetic food colors and hyperactivity in children: a double-blind challenge experiment. Pediatrics 62:975–983
Rhoads GG (1987) Reliability of diet measures as chronic disease risk factors. Am J Clin Nutr 45:1073–1079
Swanson JM, Kinsbourne M (1980) Food dyes impair performance of hyperactive children on a laboratory learning test. Science 207:1485–1487
Varley CK (1984) Attention deficit disorder (the hyperactivity syndrome): a review of selected issues. J Dev Behav Pediatr 5:254–258
Weiss B, Williams JH, Margen S et al. (1980) Behavioral responses to artificial food colors. Science 207:1487–1489
Yano K, Rhoads GG, Kagan A (1977) Coffee, alcohol, and risk of coronary heart disease among Japanese men living in Hawaii. N Engl J Med 297:405–409
Yano K, Rhoads GG, Kagan A, Tillotson J (1978) Dietary intake and the risk of coronary heart disease in Japanese men living in Hawaii. Am J Clin Nutr 31:1270–1279
Yudkin J (1964) Dietary fat and dietary sugar in relation to ischemic heart disease and diabetes. Lancet ii:4–5

Chapter 8

Principles of Behavioral Epidemiology

P. M. Vietze and M. Kiely

The title of this chapter needs some explanation.First it is not clear that there is any field called "behavioral epidemiology". Science is the systematic search for explanations. In some cases, description of the phenomena of interest is the goal. However, the search for causal mechanisms in order to understand these phenomena is the real goal of scientific study. Behavioral science and epidemiology have in common their interest in the search for causal mechanisms. The term "behavioral epidemiology" is a hybrid term which describes what we see as an emerging field. It is a term which was coined in order to define an area arising out of interest in understanding how certain etiologic factors affect a specific subset of behavioral outcomes, or how a set of health outcomes are mediated by some behavioral intervening variable (an explanatory construct which elucidates the relationship between independent and dependent variable). The outcomes in question are those usually defined by behavioral markers rather than by health or biological markers. We should say that "behavioral" is used in a generic, non-biological sense, not in the more specific sense usually reserved for "experimental analysis of behavior" or "applied behavior analysis" (Sidman 1960).

The sorts of behavioral outcomes of greatest interest are psychologic constructs such as intelligence, learning disabilities, school achievement, mental retardation, behavior disorders, psychiatric or psychologic status, adaptive behavior or some measure of language performance. The intervening variables which might be of interest could include a measure of some lifestyle factor such as how much a mother breastfed her infant, how many drinks of alcoholic beverage were consumed, child neglect or abuse, a measure of self esteem, attitudes toward some particular target or some other measure which reflects a psychologic or behavioral state. On the surface, it may appear that these outcomes or intervening variables are as easy to measure as any others. There must be records and databases in which many of these outcomes are stored and are easily accessible for use in epidemiologic studies. It is true that some of this information may be found in databases in the form of school or clinic records. The most common are achievement test records or intelligence test results. But many of these measures are not routinely collected and bring with them measurement problems not usually encountered by epidemiologists who might be interested in these outcomes. For some measures, intelligence for example, the measure is the result of an intense evaluation process which can take up to 90 minutes. While the resulting score is stable, reliable and valid, there are usually no short-cuts to measurement and no other markers which might serve as adequate surrogates. It might be desirable to have as valid or reliable a measure of memory or problem solving ability or a host of other cognitive measures. Unfortunately, they really do not exist.

The purpose of this chapter is to briefly define some of the main methodologic features of epidemiology and of behavioral science, give some examples of how these two fields have come together and finally provide some suggestions for future directions to pursue.

Epidemiology Methodology

The primary use of epidemiology is the elucidation of causal mechanisms of health disorders. It has also been used to describe and explain patterns of disorders (i.e. how the number of persons in the population with the disorder varies over time or by geographical area).

There are two main types of epidemiologic studies: cohort and case–control studies. Although hybrid study designs are used, for the purpose of simple explanation we will describe only the two main types.

Cohort Studies

Cohort studies are designed to compare rates of disorders in groups of people who differ in certain personal characteristics or exposures. The groups of people, or cohorts are observed over time to determine the rates of disorders within each group. Such studies are conducted to determine whether certain characteristics or exposures are causally related to the outcome of interest.

There are two types of cohort studies: prospective and retrospective. In a prospective cohort study the characteristics or exposure may or may not have taken place, but the disorder will not yet have occurred. Rather, the study must follow individuals over time to determine whether, and at what rate, the disorder appears. In a retrospective cohort study, both the exposure and the disorder will have already occurred when the study begins. The advantage of a retrospective study is that the cohorts can be identified from available information and the waiting period between exposure and outcome is eliminated. Retrospective cohort studies can only be conducted if sufficient information on exposure and subsequent outcome is available.

Figure 8.1 shows the design of the Dutch Famine Study, a retrospective cohort study. Toward the end of World War II, the Germans cut off food supplies to the western part of Holland. This study is unique in that data are available for a total population of 19-year-old survivors of male births in the Netherlands born during the years 1944 and 1945. Additionally, it examines acute undernutrition in a formerly well-fed population. Universal military induction examination records were linked to records of food rations during the famine and to birth and death

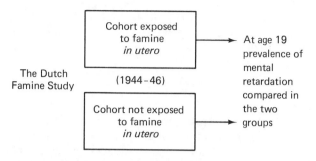

Fig. 8.1. Example of a cohort study.

records (Stein et al. 1975). Thus, this study examined a variety of outcomes among individuals exposed to famine *in utero*.

Several factors influence the selection of participants in cohort studies. Individuals who have undergone an unusual occupational or environmental exposure are prime sources of cohorts, and may be selected in order to investigate the effects of these exposures. For example, groups of pregnant women with a common history of nutritional deprivation have been studied to examine the long-term effects upon their offspring. Information on the exposure serves as the basis for identifying the cohorts. The extent of any adverse outcome on the offspring serves as the basis for identifying the risk due to exposure.

There are three main strategies for establishing comparison groups in cohort studies: a within-group comparison is actually a single large cohort within which all individuals are classified as either exposed or not exposed. This single large cohort can mitigate the need for an outside comparison group. A between-group comparison involves the establishment of a non-exposed cohort which is demographically similar to the exposed cohort. The third strategy for establishing a comparison group is the use of the population at large as the non-exposed cohort. The rate of the disorder in the exposed cohort is compared with the rate of the disorder in the population. This last strategy is limited to those circumstances in which (1) the rate of the disorder is known for the population and (2) the characteristics (e.g. demographic and geographic) of the study cohort and the population are comparable.

The analysis of a cohort study is basically a comparison of the rates of the outcome of interest (i.e. morbidity and mortality) between persons exposed and persons not exposed. It should be noted that the denominator of the rates is expressed as a function of both cohort size and time (i.e. person-years) because longitudinal studies lose subjects over time and must also account for the varying time periods in which different subjects enter and remain in the study.

There are three risk estimates which are commonly used as a measure of association between exposure and outcome. Relative risk refers to the ratio of the rate of the disorder among the non-exposed group to the rate of disorder among the exposed group. The relative risk provides a direct measure of the risk due to an exposure. Attributable risk refers to the rate of disease in exposed individuals that can be attributed to the exposure. Population attributable risk fraction refers to the proportion of all cases in a defined population which can be directly attributed to an exposure or characteristic. This measure provides an estimate of the impact that a specific exposure or characteristic may have on the outcome in the population. The population attributable risk fraction also allows for an estimation of the relative benefit to the population of reducing or eliminating the exposure (MacMahon and Pugh 1970).

Case–Control Studies

Case–control studies are designed to contrast the rate of characteristics or exposures among persons who either manifest or do not manifest a given disorder. Individuals considered to have the disorder are defined as cases, and individuals who are considered not to have the disorder are defined as controls. Cases and controls are compared to determine characteristics or exposures which are more common to one group and are relevant to the hypothesized etiology of the disorder. Case–control studies are most useful in testing hypotheses about etiologic factors associated with a specific disorder.

As the purpose of case–control studies is to identify characteristics or exposures associated with a disorder, precise diagnostic criteria are crucial. Ideally, there should be no ambiguity about the diagnosis of cases or controls. The erroneous identification of cases or controls could mask valid group differences.

Figure 8.2 shows the design of a case–control study of diet and colorectal cancer in African-Americans. Cases were all black patients hospitalized for the first time with colorectal adenocarcinoma at hospitals participating in the California Tumor Registry. For each case, two black hospital controls and one healthy black control, having a check-up, were interviewed. Controls were free of gastrointestinal tract cancer and were matched to the case by sex, age within 5 years, and hospital. Participants were questioned on their consumption of a variety of food items. This study examined diet as an exposure factor among patients with colorectal cancer (i.e. cases) and among other hospital patients and health facility outpatients (i.e. controls) (Dales et al. 1979).

Cases are generally recruited for case–control studies from specific target populations, such as cases in treatment, or from the population at large. Selecting cases from existing registries or treatment programs is the most efficient strategy.

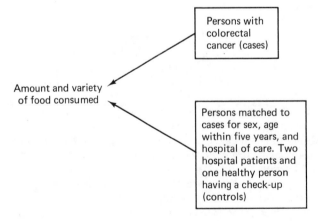

Fig. 8.2. Example of a case–control study.

However, such selection procedures may result in a non-representative group of cases. The cases may be unrepresentative in terms of the severity of their disorder or their personal characteristics (i.e. age, sex, race). Selecting cases from the population will increase the difficulty and expense of case-finding, but such efforts will enable researchers to estimate the rates and severity of the disorder in the population.

The purpose of controls is to determine whether the frequency of the exposure or characteristic differs between cases and a comparable group without the disorder. If the frequency of the characteristic or exposure differs between cases and controls, then one can assess the contribution of the hypothesized etiologic factor to the manifestation of the disorder. If the frequency of the characteristic or exposure does not vary between cases and controls, then it is unlikely that those factors are causally associated with the disorder.

There are several factors to be considered in defining a control group. One must be careful to obtain all information on personal characteristics or exposures in a similar manner for cases and controls. Biases in collecting such information could potentially mask or lead to spurious group differences. The researcher must be careful to insure the validity and reliability of respondent recall.

Cases and controls should be drawn from the same or similar populations. In selecting a control group it is important to consider the associated economic and logistic costs. As a minimum, controls should be representative of the population from which they are drawn, or the manner in which they are not representative should be known. There are several sources of controls for case–control studies. Subjects from the same medical facility as the cases may be a source for controls. This source is often the most practical because access to a database has already been established and the quality of the information is likely to be comparable for both cases and controls. However, this source of controls may be unrepresentative of the population at large, making it difficult to extrapolate from any study findings. When control groups are selected from a clinical facility, it is important that they are drawn from a range of diagnostic categories, rather than a single category. If a single diagnostic group is chosen for the control group and group differences are found, it will be difficult to determine whether it is the cases or controls who differ from the general population.

Family members may also be used as a source of controls. The advantage of family members as controls is that they will be likely to have the same ethnic and social background as the cases. However, family controls are inappropriate when the factor being studied is common to both the cases and their family members (e.g. environmental exposures). Another source for controls may be associates of cases (e.g. neighbors, classmates, fellow employees). It can be especially time consuming to recruit associates of cases and collect the necessary information.

In addition to selecting an appropriate control group, the researcher must decide on the number of persons within the control group(s). When the cost of obtaining data is high and there are a potentially unlimited number of cases, the best choice is to have equal numbers of cases and controls. If the number of cases is limited or if the cost of obtaining data is low, then the best choice is to have multiple controls. By selecting more controls than cases, the statistical power of the study can be increased; however, minimal power is gained if there are more than four controls for each case (Gail et al. 1976).

The sampling of controls for a case–control study can be either systematic or random. A systematic sample is one in which the entire list of potential controls is ordered in terms of one or more characteristics and the controls are then systematically selected (i.e. every nth control). When systematic sampling is used, the order of the list of potential controls should not be arranged so that it varies on any of the factors of interest. Systematic sampling is done to insure a representative distribution in the control group. Another strategy for selecting controls involves the pair-wise matching of controls to individual cases. The goal of matching procedures is to insure that cases and controls are similar on variables which may have an effect on the dependent variable and also vary with the independent variable (Susser 1973). The most common matching characteristics are age, sex, and ethnicity (Mausner and Kramer 1984). Failure to match cases and controls on the necessary characteristics may lead to spurious results. However, it may become difficult to assemble a control group if too many matching criteria are chosen.

The analysis of case–control studies is a comparison of the frequency of the suspected etiologic factors among cases with the frequency of the same factor among controls. The comparison of cases and controls is the odds ratio, and provides an estimate of the "true" relative risk. The likelihood of demonstrating actual differences between cases and controls is increased in subjects who are classified across multiple categories. For example, the relationship between alcohol consumption and fetal alcohol syndrome can be assessed by contrasting infants born to mothers who consumed alcohol during gestation with infants born to mothers who abstained from alcohol (Russell 1977; Olegard et al. 1979). The hypothesized relationship would be easier to demonstrate if drinking mothers were classified according to their quantity or frequency of drinking. If a dose–response pattern were demonstrated as a function of alcohol consumption, then more strength could be given to the hypothesis that prenatal alcohol exposure leads to fetal alcohol syndrome (MacMahon and Pugh 1970).

Establishing Cause

Much of epidemiologic research is concerned with establishing the presence and direction of causal relationships among variables. Susser, in 1973, described criteria which can be used to determine whether an association is causal (Susser 1973).

Consistency. This criterion requires that similar findings are obtained when studies are performed under varying conditions (e.g. different populations or study designs). For example, two studies investigated alcohol consumption and spontaneous abortion and found consistent results. One addressed the problem prospectively (Harlap and Shiono 1980), the other retrospectively (Kline et al. 1980). The two studies are consistent in that they found excessive drinking increased the risk of spontaneous abortion.

It should be noted that if methodologic errors are repeated between studies, the results may be consistent but spurious.

Strength of association. The degrees to which variables occur together is a measure of their strength of association. The stronger the association, the more probable that the association is causal. The relative risk and the odds ratio each serve as an index of the strength of association in epidemiologic research. For example, Ellenberg and Nelson (1979) reported rates of cerebral palsy by birthweight. The incidence of cerebral palsy increased with decreasing birthweight. The strength of the association between low birthweight and cerebral palsy was demonstrated by a steep rise in the relative risk as the birthweight decreased (see Table 8.1).

Table 8.1. The incidence of cerebral palsy increases with decreasing birthweight

Birthweight (g)	Cerebral palsy cases per 1000	Relative risk[a]
⩾2501	3.70	1.00
1501–2500	13.91	3.76
⩽1500	90.43	24.43

From Ellenberg and Nelson (1979).
[a]Compared with the group with birthweight ⩾2501 g.

Time. In order for a factor to be considered causally related to a disorder, exposure to the factor must precede the manifestation of the disorder. The time sequence between prenatal exposure to infectious diseases and postnatal outcome is relatively simple to describe. However, with chronic diseases having an insidious onset (e.g. multiple sclerosis), the time lapse between a hypothesized exposure and outcome is often difficult to establish.

Specificity. This criterion of judgment refers to the extent to which the occurrence of one variable predicts the occurrence of another variable. The highest level of specificity is when a certain event always predicts the occurrence of a disorder and the disorder occurs only after that event. An example of complete specificity is an extra 21st chromosome and the constellation of characteristics associated with Down syndrome. A high level of specificity is one indication that the variable being studied is causally associated with the observed effect. However, it is important to remember that even though high specificity increases the likelihood that a relationship is causal, it does not prove the causal association.

Coherence. The criterion of coherence considers the credibility of the association of two variables. An association between two variables is considered coherent if the relationship is plausible in terms of current scientific knowledge.

Reliability and Validity

When designing a study, the researcher must be sure that instruments are both reliable and valid. A reliable questionnaire is one that collects information that is replicable; checks on repeatability assure reliability. One approach to assessing reliability is to reinterview a sample of the respondents and compare responses from the two interviews. Using the same interviewer holds constant the effect of variability among interviewers and measures consistency of the respondents' recall. Using different interviewers for the two interviews measures consistency of recall as well as variation in response to different interviewers. Another method to assess reliability is to repeat questions in different forms within the same interview. This is called internal consistency.

A valid questionnaire measures "that which it is purported to measure." Bennett and Ritchie (1975) divide validity into three components:

1. Relevance: Does the questionnaire obtain the information it was designed to retrieve?
2. Completeness: Does the questionnaire obtain all the desired information?
3. Accuracy: Do the responses reflect reality?

One approach to assessing validity involves confirming the obtained information from supplemental sources. It should be recognized that, strictly speaking, what the researcher is measuring is consistency between the interviewer and the supplemental sources since the supplemental source can also be incorrect. An additional method for checking validity of the research instrument is to include items to test internal validity. For example, information could be obtained on drug usage, including the use of drugs directly relevant to the study disease as well as drugs that are known to be irrelevant to the study disease. Systematic bias can be detected since cases and controls should have comparable drug usage for the irrelevant drugs, but should show differences in the drugs of interest. The last method to assess validity is to ask questions that have only one reasonable answer. It is also important to guard against having questions worded in such a way that norms of propriety and social desirability are not violated. The "demand characteristics" of the questions must be considered. This means that the question cannot suggest only one answer and the question cannot be biased in only one direction.

Behavior Science Methodology

Although it has been possible to provide an overview of the methodologic features of epidemiology in a fairly concise manner, the same is not true for the other half of our amalgamation. One reason is that behavioral scientists merely think of methodology and research design as basic to research. There are a number of reasons for this. One may be because many of the phenomena studied by behavioral scientists are often available to the researcher only through self-report and thus appear to be subjective. Therefore, behavioral scientists have become especially sensitized to the threats to validity encountered in such study and have devised ways to insure that all measures are reliable and valid. By including a variety of forms for obtaining self-report measures, establishing validity of self-report measures prior to their use as outcome variables, and insuring test–retest and internal consistency forms of measurement reliability, the subjective nature of many measures becomes unimportant.

In general, most behavioral scientists see experimental study with random assignment of subjects to treatments as the method of choice when possible. However, random assignment of subjects to treatments may not always be possible in the real world or field studies which behavioral scientists conduct. In the absence of being able to exert experimental control over the phenomena under study, behavioral scientists have looked to other ways for controlling error variance in their research.

The behavioral scientists who conduct experimental analysis of behavior studies have developed techniques to study single subjects. Using their techniques, which are based on principles of learning, they systematically apply and withdraw experimental treatments and observe the results in their subjects. By varying conditions systematically and by studying these conditions over long periods of time, they are able to infer the effects of these conditions on outcomes of interest. These techniques work well for independent or treatment variables which the scientists can control. However, in the case of phenomena where there is no possibility for controlling treatments, such experimental studies are not possible.

Such studies include those in which the subjects of interest are people or animals for which a particular treatment could not ethically be applied and where subjects cannot be randomized to treatments. These are, then, primarily observational studies of exposures to conditions which cannot be experimentally manipulated or induced and where the outcomes of interest are primarily behavioral. The four sets of factors which must be considered in behavioral epidemiology studies are threats to statistical conclusion validity, threats to internal validity, threats to construct validity of putative causes and effects, and threats to external validity. This characterization is one developed by Cook and Campbell (1979) for what they call "quasi-experimentation" and is an elaboration of an earlier analysis by Campbell and Stanley (1963). The model for this conceptualization is experimental manipulation where field studies often preclude such manipulation. The notion of quasi-experimentation is appropriately applied to "behavioral epidemiology" in which an event occurs and its effect on some behavioral outcome is of interest.

Statistical Conclusion Validity

Statistical conclusion validity concerns the relationship between two variables of interest. It asks the question, "Is there a relationship between the independent and dependent variables?" Included in this issue are the questions of statistical power, whether there is covariation and the strength of the relationship. There are a number of threats to this sort of validity including low statistical power, violated assumptions of statistical tests, error rate, reliability of measures (test–retest or stability), reliability of treatment implementation, random irrelevancies in the setting and random heterogeneity of respondents.

Internal Validity

Internal validity is concerned with the causal relationship between the two variables of interest and the direction of effect. Threats to internal validity include history, maturation, testing, instrumentation, statistical regression, selection, mortality, interactions with selection, diffusion or imitation of treatments, compensatory equalization of treatments, compensatory rivalry by respondents receiving less desirable treatments, resentful demoralization of respondents receiving less desirable treatments and ambiguity about the direction of causal influence. Random assignment of subjects to treatments rules out most of these threats. However, the last five threats cannot be reduced by randomization and must be ruled out individually according to the circumstances of the study. In field studies, where randomization is not possible, each threat to internal validity must be considered and eliminated in order to be able to draw the correct conclusion about the outcome. Some of these threats apply to studies which are experiments or quasi-experiments.

Construct Validity of Putative Causes and Effects

Construct validity of putative causes and effects is sometimes known as confounding. This occurs when a variable is represented as the presumed treatment or outcome and may be masking another variable. One example of this would be the case where a nutritional supplement is given by a supportive person as a treatment and leads to some desirable outcome. It is conceivable that the outcome may be the result of the supportive person or the actual act of giving something rather than the nutritional substance itself. Without providing a placebo and a double-blind experiment, accurate causal attribution is not possible. Among the threats to construct validity of putative causes and effects are inadequate preoperational explication of constructs, mono-operation bias, mono-method bias, hypothesis-guessing within experimental conditions, evaluation apprehensions, experimenter expectancies, confounding constructs and levels of constructs, interaction of different treatments, interaction of testing and treatment, and restricted generalizability across constructs. There are a number of ways of reducing these threats. Many of them involve elaborate experimental designs. Some of them are eliminated as a result of conducting several studies. Thus a program of research, through its systematic study of different variables over time, will solve these threats to construct validity.

External Validity

External validity is concerned with generalizability of results across persons, settings, and occasions. Threats to external validity include interaction of selection with treatment, interaction of setting and treatment, and interaction of history and treatment. Elimination of these threats involves sampling in a manner which enhances generalizability. Replication across different samples, settings and times can go a long way to insuring generalizability.

Behavioral Epidemiology Research Studies

Having outlined methodologic approaches to epidemiologic and behavioral science research some examples of studies which might be considered behavioral epidemiology research are now considered.

National Health Interview Survey: Child Health Supplement

The National Health Interview Survey is a continuous health survey conducted by the National Center for Health Statistics of the US Public Health Service in collaboration with the US Bureau of the Census. It is a national survey of 60 000 persons conducted in order to report on the health of the nation. In 1981, a section called the Child Health Supplement was added in order to provide detailed information on the children in the households being surveyed. Although this was a detailed and well-constructed interview some of the items of interest included health and behavior conditions which might be considered to have negative social stigma. For example, among a long list of conditions were included mental retardation and learning disabilities. The respondent was expected to indicate if the index child had any of these conditions. No other corroborating questions were asked, regardless of how the respondent answered. It is likely that the prevalence rate of mental retardation based on this question is underestimated. It is not clear that anyone would readily admit that his or her child has mental retardation or any other disorder which has a social stigma attached to it. This is borne out by the finding that only 0.1% of the sample had a child with mental retardation. It is clear that one of the issues which is ignored in many interview and self-report surveys is the validity of the responses. Without extensive cross-checking questions, items which place heavy demands for socially desirable responses, called demand characteristics in the social psychology literature, may yield misleading results. This is an example of how the instrument used to measure an outcome may be unreliable. Unreliable measurements are surely a threat to internal validity. The field of social psychology has produced countless studies indicating that response sets such as the social desirability of particular responses, positive or negative response bias and other problems in the construction of instruments may preclude accurate measurement and hence may produce invalid results. Accurate measurement of underlying constructs is necessary for valid conclusions. The use of single questions to assess a particular construct may also be an example of mono-operation bias. This points up the importance of extensive preliminary development of instruments and their validity.

National Longitudinal Survey of Youth

The National Longitudinal Survey of Youth originated in 1979. Its original intention was to examine the ways in which young people entered the labor market. It was initiated by the US Department of Labor and has continued with major funding from that agency supplemented with funds from the National Institute of Child Health and Human Development. This study has been continued annually through 1986 with over 10 000 respondents interviewed. The original respondents ranged in age from 14 to 21 years. Over time they have entered their reproductive years and by 1986 2918 women in the sample had produced over 5000 children. As the 1986 survey was planned, the Center for Human Resource Research at the Ohio State University, which had been conducting the survey since its initiation, began to realize that with such a large number of offspring, the survey now represented a valuable database. Since the women had reached childbearing age, questions about prenatal care, childbirth and child care had been added. Now it became evident that this database could answer questions about the etiology of many childhood events and experiences if it were maintained.

A number of behavioral scientists became very interested in having the children themselves serve as respondents. However, children are not reliable respondents on most topics until they reach the age of 8 or 10. Interest was shown in the parent–child transmission of many aspects of child development. In order to carry this out it was evident that the children would have to be tested using tests from the armamentaria of child psychology and child development. A large panel of child development experts was convened to plan the protocol and a test battery designed. It was soon discovered that the time required for the test battery was too long. A standardized IQ test takes at least an hour to administer and the cost for this survey made it possible for the child testing to last no more than 1 hour. As a result of this practical limitation, short forms of the tests needed to measure the cognitive, motivational and temperament outcomes of interest would have been desirable. Unfortunately, such short forms were not available for most measures. Instead, subscales with reported good validity and some measures constructed by some of the consultants were used. The list of measures used may be found in Table 8.2. It should be noted that, as Zill has stated (Zill et al. 1984), it is important for child development behavior scientists to develop valid and reliable markers of child development; these would be comparable to health markers, which have been demonstrated to represent aspects of child and community health such as birthweight, infant mortality, ponderal index and others.

Table 8.2. Child assessments: National Longitudinal Study of Youth

Home observation for measurement of the environment
Body parts
Peabody Picture Vocabulary Test (PPVT)
Memory for location
McCarthy Scale of Children's Abilities: verbal memory subscale (3 years–6 years, 11 months)
Wechsler Intelligence Scale for Children: revised digit span subscale (7 years and older)
Peabody Individual Achievement Test (PIAT): math subscale (5 years and older)
Peabody Individual Achievement Test (PIAT): reading recognition subscale (5 years and older)
Peabody Individual Achievement Test (PIAT): reading comprehension subscale (5 years and older)

Defective Infant Formula Study

In the late 1970s, an apparent epidemic of failure to thrive was reported among infants throughout the USA. After a period of time, it was discovered that most of these infants had been given one of two types of soy-based infant formula. Upon analysis, the formulas were found to be missing chloride, necessary for normal growth. Failure to thrive was especially prevalent in cases where the formula was the exclusive food for the infants and the symptoms presented were chloride depletion or chloride deficiency syndrome. The Centers for Disease Control (CDC) established a registry of infants with this syndrome who had been given the defective formula. Later these children were diagnosed as having hypochloremic metabolic alkalosis. A study of children who had been exposed to this defective formula was begun at the National Institutes of Health (NIH). The goal of the study was to discover the sequelae of the defective formula. Among the outcomes included in the protocol was a full-scale measure of infant behavioral, cognitive and motor development. Findings from this study were reported recently by a team of NIH researchers and indicated a dose–response relationship between length of exclusive exposure to the defective formula and measures of infant behavioral development at age 2, and in mental and motor functioning at age 4. Unfortunately, there was some attrition between the initial evaluations and the 4-year follow-up. There were also no controls in this study, save a few sibling controls. Since the sample size was small, the even smaller number of siblings tested could not be used for any comparisons.

The NIH was directed by Congress to conduct a study to determine the consequences of ingestion of the defective formula. This study has been a collaboration between epidemiologists and behavioral scientists. In fact, rather than one study, a series of studies has been conducted in order to determine the effect of the chloride deficient infant formula on child behavior and development. The resulting series of studies, each of which answers a slightly different question, has had to make compromises in methodology in order to accommodate the different approaches used by epidemiologic and behavioral researchers. The first study, described above (Willoughby et al. 1987), was a case-only study with no adequate controls.

The second study was a retrospective cross-sectional study. In this study an attempt was made to eliminate ascertainment bias by taking a population sample in a small school district, identifying cases as those children who had been given soy-based formula as infants and dividing this group into those who had had the defective formula and those who had had other soy formulas. All of these children were tested using a standard psychometric instrument. In order to evaluate other possible independent variables, information on parent education, race, and income was also collected. The results supported the earlier findings from the case study, only this time the children were age 6 (A. Willoughby, in preparation). Again, there was a significant deficit in some areas of cognitive and motor functioning in the group who had been given the defective formula.

This study has served as a pilot study for a more extensive replication in a larger population that will have sufficient power to detect meaningful differences and thus avoid the threat to statistical conclusion validity. This study utilizes the same method as the one just described except that the population base is large enough to select multiple controls for each case of a child who had been given defective formula. While incomplete, this study has avoided the problem of ascertainment bias which may plague epidemiologic studies. The risk it does run is that the retrospective recall of what formula the child was given may be inaccurate. In addition, the outcome measures used are limited in that only a standardized intelligence test and one short language test were used. The selection of instruments is based on our earlier findings and anecdotal reports in the literature and from a handful of clinicians who had experience with children who had been exposed to the defective formula. Were there better outcome measures, the results of this study and the others conducted in this series might have greater precision in understanding the effects on development of the chloride deficient formula.

The third study is a longitudinal follow-up study of children who were identified as having hypochloremic metabolic alkalosis prior to the publicity which appeared in the news media during the period when the formula was on the market. The CDC registry was established before the publicity campaign and before the products were removed from the market by the Food and Drug Administration (FDA). Therefore, despite the fact that there were almost 150 names on the registry, only children whose names were reported prior to the public announcements are eligible for this study. The study consists of a medical examination, neurologic examination, anthropometry and an extensive battery of tests measuring cognitive, neuropsychologic, auditory and language functions. Controls are drawn from the school-based study described above and also from the families of the index cases. It is expected that the results of this third study will provide some insights into the long-term consequences of chloride deficient infant formula for biological and behavioral development. However, the large effort involved in obtaining the data suggests that the development of outcome instruments which are valid and reliable and which can be administered easily in the field, would have permitted this study to have taken place earlier in the lives of these children and with greater ease.

Behavioral Outcome Measures

One of the themes in the descriptions and discussion above is that the outcome measures used in behavioral epidemiology studies have some limitations. In many cases, we are interested in how exposure to a health risk affects intelligence or some other behavioral outcome or developmental marker. One of the reasons why intelligence is used so often as the outcome measure, is that we know how to measure it reliably. Intelligence tests can be administered very reliably. The test scores are also considered to have excellent validity. However, they may not always provide answers to the right questions. They also may provide only summary measures and not provide adequate information about mental processes and processing ability. A number of psychologists have discussed this extensively in recent years, e.g. Vietze and Coates (1986) and Zelazo (1988). Intelligence tests were developed to evaluate intelligence and not to be used as evaluation or outcome measures in themselves. They often cannot be used with the populations in epidemiologic studies since they were not standardized to the populations being studied. Finally, their strength, the reliability of measurement, may be the result of the fact that they are norm-referenced tests rather than criterion-referenced tests. Many of these arguments have been made in other contexts such as in the limitations of standardized instruments when applied to minority populations. The development of widely accepted behavioral outcome measures, or behavioral and developmental indicators, could provide solutions to many of these problems. This is similar to the suggestion made by Zill et al (1983) for the development of childhood social indicators. However, Zill and associates focus on health and school indicators and it is suggested that behavioral indicators in other settings might be needed. It is recommended that outcome measures which have validity for the most typical behavioral and developmental criteria be developed so that epidemiologic research might become more meaningful.

Summary and Conclusions

We have described a field called "behavioral epidemiology" which is an amalgamation of epidemiology and behavioral science. It consists of research in which relationships between exposures and behavioral outcomes are studied. A brief description of epidemiologic methodology and of behavioral epidemiology has been given. Several examples of behavioral epidemiologic studies have been described and some critical comments made in order to illustrate some problems in behavioral epidemiology. Finally, a recommendation is made that behavioral outcome measures be developed which provide cost-effective ways for measuring behavioral outcomes which have relevance for epidemiologic questions and behavioral development.

References

Bennett AE, Ritchie K (1975) Questionnaires in medicine: a guide to their design and use. Oxford University Press, London

Campbell DT, Stanley JC (1963) Experimental and quasi-experimental designs for research. Rand McNally, Chicago

Cook TD, Campbell DT (1979) Quasi-experimentation: design and analysis issues for field settings. Rand McNally, Chicago

Dales LG, Friedman GD, Uri HK, Grossman S, Williams SR (1979) A case-control study of relationships of diet and other traits to colorectal cancer in American Blacks. Am J Epidemiol 109:132–144

Ellenberg JH, Nelson KB (1979) Birth weight and gestational age in children with cerebral palsy or seizure disorders. Am J Dis Child 133:1044–1048

Gail M, Williams R, Byar DP, Brown C (1976) How many controls? J Chron Dis 29:723–731

Harlap S, Shiono PH (1980) Alcohol, smoking and incidence of spontaneous abortions in the first and second trimester. Lancet ii:173–176

Kline JK, Stein ZA, Shrout P, Susser MW, Warburton D (1980) Drinking during pregnancy and spontaneous abortion. Lancet ii:176–180

MacMahon B, Pugh TF (1970) Epidemiology: principles and methods. Little, Brown, Boston

Mausner JS, Kramer S (1984) Epidemiology: an introductory text, 2nd edn. Saunders, Philadelphia

Olegard R, Sabel DG, Aronsson M et al. (1979) Effects on the child of alcohol abuse during pregnancy. Acta Paediatr Scand [Suppl] 275:112–121

Russell M (1977) Intra-uterine growth in infants born to women with alcohol-related psychiatric disorders. Alcohol Clin Exp Res 125:225–231

Sidman M (1960) Tactics of scientific research. Basic Books, New York

Stein Z, Susser M, Saenger G, Marolla F (1975) Famine and human development: the Dutch hunger winter of 1944–1945. Oxford University Press, New York

Susser M (1973) Causal thinking in the health sciences: concepts and strategies in epidemiology. Oxford University Press, New York

Vietze PM, Coates DL (1986) Using information processing strategies for early identification of mental retardation. TECSE 6:72–85

Willoughby A, Moss HA, Hubbard VS et al (1987) Developmental outcome in children exposed to chloride-deficient formula. Pediatrics 79:851–857

Zelazo PR (1988) An information processing paradigm for infant-toddler mental assessment. In: Vietze PM, Vaughan HG (eds) Early identification of infants with developmental disabilities. Grune and Stratton, Phiiladelphia, pp 299–317

Zill N, Sigal H, Brim OG (1983) Development of childhood social indicators. In: Zigler E, Kagan SL, Klugman E (eds) Children, families and government. Cambridge University Press, New York, pp 188–222

Zill N, Peterson JL, Moore KM (1984) Improving national statistics on children, youth and families. Child Trends Inc, Washington, DC

Chapter 9

Identifying the Genetic Component of Dietary–Behavioral Interactions: A Challenge for Genetic Epidemiology

R. H. Ward

Jack Sprat could eat no fat;
his wife could eat no lean
And so between them both,
they kept the platter clean

English nursery rhyme

One feeds on lard and yet is leane
And I but feasting on a beane
Grow fat and smooth: The reason is
Jove prospers my meat more than his

Robert Herrick 1591–1674

Preamble

Throughout history there has been an intuitive recognition that a person's "intrinsic constitution" may modify the relationship between dietary profiles and their consequences. The two rhymes above demonstrate the age-old perception that "intrinsic constitution" can influence the consequences of dietary profiles in one of two distinct ways: metabolic and behavioral. The nursery rhyme suggests that striking differences in dietary preferences (and their resulting consequences for obesity) may be due to stronger imperatives than mere idiosyncratic choice. In constrast, Herrick's rhyme stresses that metabolic differences, rather than differences in dietary intake, may underlie the physiologic consequences leading to obesity. While the contemporary perception of the interaction between diet and behavior is couched in more sophisticated terms, there is increasing evidence that this interaction is influenced by the action of genetic factors on both major causal pathways: metabolism and neurophysiology. Consequently, as summarized in Velazquez and Bourges (1984), there is a plethora of important research issues that require the joint effort of geneticists and nutritionists for their resolution. Despite the recognition of these problems, progress in addressing genetic issues has been slow, in large part because of the difficulty in defining testable hypotheses.

A Conceptual Paradigm

In order to evaluate the genetic influence on phenomena as complex as dietary–behavioral interactions, the definition of a plausible paradigm of causality is an important first step. Without a conceptual model of causation, it will be extremely difficult to apply the techniques of genetic epidemiology. The establishment of a set of probable causal pathways leads to a set of testable hypotheses and these in turn define the most appropriate study design and analytic strategies that should be employed. Figure 9.1 represents one such conceptual paradigm in which genotypic differences influence the dietary–behavioral interaction by virtue of their effect on either of the two main causal pathways that lead to an individual's "intrinsic constitution." Here the basic interaction between diet and behavior is presumed to be dyadic: diet influences behavior and behavior influences diet:

As indicated by Fig. 9.1, genetic and environmental pathways can influence either direction of this interaction. Further, just as influences operate through two biologically distinct pathways – metabolic and neurophysiologic – so the distribution of environmental influences can be subdivided into two components: ecologic and sociocultural. In this context, ecologic factors include economic and demographic constraints, as well as the distribution of food, while sociocultural factors include the social perception of body form, in addition to the cultural imperatives that underlie food and/or nutritional preferences. Hence, the great variety of genetic and environmental factors that influence dietary–behavioral interactions can be grouped into these four distinct categories of causal influence. This implies that the health-related consequences of specific dietary–behavioral interactions can be mediated by both genetic and environmental factors, in most cases operating through the metabolic and sociocultural pathways. Obesity with its pervasive and extensive set of morbid consequences (Forman et al 1986) exemplifies the complex nature of such interactions, since both categories of genetically influenced pathways (metabolic and neurophysiologic) are causally

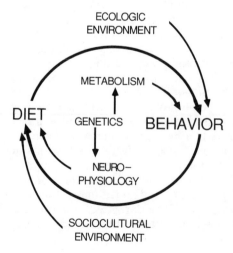

Fig. 9.1. Conceptual paradigm of the causal pathways leading to dietary–behavioral interactions. Genetic influences on causation are deemed to act primarily through metabolic or neurophysiologic intermediates, while environmental influences may be ecologic or sociocultural.

implicated in the development of obesity, as are both types of environmental influence (ecologic and sociocultural).

An important corollary of the paradigm illustrated by Fig. 9.1 is that it posits that the two major intrinsic categories (metabolic vs. neurophysiologic) are likely to influence different sides of the fundamentally dyadic interaction. Hence, genetic factors are most likely to influence the "diet-to-behavior" interaction by way of underlying differences in metabolic profiles. A simple example is afforded by the effect of genetic variants of liver aldehyde dehydrogenase which result in an ethanol flush reaction after the consumption of alcohol (Bosron and Li 1986). The discomfort caused by this reaction tends to modify behavior by reducing alcohol intake. The consequences of lactose intolerance (see below) is another example of how genetic dietary behavior is affected by differences in metabolism.

Conversely, the influence of genetic factors on the behavior-to-diet interaction is most likely to occur via differences in neurophysiology. Alcohol use provides an illuminating example. A number of extensive genetic studies (discussed below) indicate that many people who abuse alcohol do so because of genetically derived differences in monoaminergic profiles which result in personality profiles that exhibit a propensity for alcohol abuse (Cloninger 1987; Cloninger et al. 1988). In analogous fashion, a case can be made for asserting that genetic constitution is an important determinant in many of the eating disorders (e.g. anorexia nervosa, bulimia), which represent extreme forms of the dietary–behavioral interaction. For example, the psychologic profile associated with bulimia (Abraham and Beaumont 1982) is likely to have its origin in genetically defined neurophysiologic profiles. Similarly, it is plausible that genetic differences in metabolism underlie the characteristic taste preferences exhibited by bulimics (Drewnowski et al. 1987).

The contrast between the influence of ecologic and sociocultural factors on different directions of the dyadic dietary–behavioral interaction is less apparent. However, it seems likely that sociocultural factors, rather than ecologic factors, influence the diet-to-behavior interaction. For example, perceptions of obesity not only vary widely from one culture to another but, in the context of increasing Westernization, can have a considerable effect on dietary profiles and their consequences (Takahashi 1986). These sociocultural perceptions are equally important in fully Westernized countries and in adolescents interact with their own level of obesity resulting in altered dietary preferences (Worsley et al. 1984).

Food availability is perhaps the most obvious way in which ecologic factors influence the behavior-to-diet interaction. In general, the influence of such ecologic parameters is most readily perceived in small scale, traditional, societies such as the Tarahumara of Mexico, whose sparse environment leads to extremely low levels of fat intake and hence to lowered cholesterol levels (McMurray et al. 1982). However, similar ecologic factors also operate in modern day America – though less dramatically – with the result that the characteristic distribution of obesity by race and socioeconomic status can also be interpreted in ecologic terms (Forman et al. 1986).

These distinctions are somewhat arbitrary and are certainly not exclusive; sociocultural factors and metabolic profiles certainly exert some influence on the behavior-to-diet interaction, just as neurophysiology and ecology play a role in the diet-to-behavior interaction. Nevertheless, it appears that distinct combinations of genetic and environmental factors operate to influence the respective directions of the dietary–behavioral interaction. In addition, as indicated by Fig.

9.1, the dietary–behavioral interaction is dynamic, such that each community will develop a specific dietary–behavioral profile that reflects the genetic and cultural milieu. Any modification in the distribution of either genetic factors or environmental factors will likely result in the establishment of a new dietary–behavioral profile. Hence, neither causal factor can be ignored.

One important consequence of the paradigm illustrated in Fig. 9.1 is that the impact of genetic, or environmental, influences on the dietary–behavioral interaction can result in the establishment of feedback loops which may be either negative or positive. In general, negative feedback loops will tend to be beneficial since they act to increase homeostasis. Conversely, positive feedback loops tend to be detrimental to health, since they result in disruption of homeostatic situations.

Negative Feedback Loops: Beneficial Effects

An example of a negative feedback loop with beneficial effects is given in Fig. 9.2. This illustrates the general situation where genetic or environmental factors influence behavioral and/or dietary strategies in such a way that the intake of deleterious foods is diminished, resulting in adaptive homeostasis. While there is extensive literature on the relationship between cultural practices and the avoidance of harmful foods, it should be stressed that genetic factors can also play a key role in the process. Informative examples of the manner in which genetic factors can accentuate beneficial negative feedback loops involve genetic differences in metabolism.

The restriction of alcohol intake due to the occurrence of the ethanol flush reaction in individuals with genetically different liver aldehyde dehydrogenases (Bosron and Li 1986) represents an example of a beneficial negative feedback loop. Another, even more important, example with considerable public health importance is the phenomenon of lactose intolerance. This recessively inherited condition stems from a deficiency of lactase activity in the brush border of the cells of the small intestine resulting in an inability to absorb lactose. As the

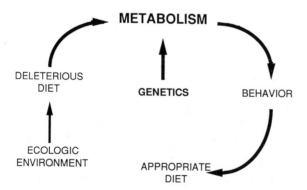

Fig. 9.2. Diagram of a negative feedback loop with beneficial effect. In this particular paradigm, genetic differences in metabolism lead to modification of behavior that culminates in avoidance of deleterious foods. Examples of such phenomena include the alcohol flush reaction and lactose intolerance.

failure to absorb lactose results in considerable abdominal discomfort after drinking milk, individuals with lactose deficiency express their "lactose intolerance" by avoiding milk and other dairy products. Since the deficiency in lactase activity is only manifest in adults, and children well past weaning, it has been argued that the recessive phenotype of lactose intolerance represents the norm for our species, and that the dominant phenotype of lactose tolerance only arose after populations became reliant on domesticated cattle (McCracken 1971). Certainly lactose intolerance is widespread and, with the exception of Caucasians and certain groups in East Africa (who herd cattle), it represents the physiologic norm for most populations, especially those in developing countries (Simoons 1978).

An important public health consequence of lactose intolerance results from the fact that affected individuals have a very different milk intake compared with the norm in Caucasian populations (Lisker et al. 1978). While this behavioral modification is beneficial for the individual, the widespread distribution of lactose intolerance in developing countries poses significant problems for implementing dairy-based food supplementation programs – a fact which is still insufficiently appreciated. It should also be noted that the degree with which lactose intolerance is expressed in individuals with lactase deficiency appears to be influenced by major differences in socioeconomic strata (Lisker et al. 1980). Whether this is mediated by ecologic or sociocultural factors is not yet understood. A variety of other beneficial negative feedback loops due to underlying genetic differences in metabolism have been identified, such as the reduction of sugar intake by individuals with fructose intolerance, with a corresponding reduction in dental caries. However, few others appear to have the global distribution and public health importance of lactose intolerance.

In theory, beneficial feedback loops could also result from genetically determined metabolic differences leading to behavioral strategies which will increase the intake of scarce foods that are required for optimal physiologic well-being. While the ethnographic literature abounds with examples where cultural practices appear to enhance the intake of scarce nutrients, it is difficult to identify convincing examples where this type of feedback loop is accentuated by underlying genetic differences in metabolism, or physiology. Similarly, there are few examples where genetic differences in neurophysiology mediate behaviors which accentuate beneficial negative feedback loops. As indicated below, behavioral differences mediated by underlying neurophysiologic profiles seem to be predominantly associated with positive feedback loops with a detrimental impact on the individual.

Positive Feedback Loops: Detrimental Effects

A positive feedback loop occurs when the interaction between diet and behavior exacerbates the effect of an outside stimulus, as opposed to a negative feedback loop which tends to diminish the effect of an external perturbation. If the external stimulus is deleterious the action of the positive feedback loop will result in a departure from an existing physiologic optimum, leading to what Scriver (1984) labels "dishomeostasis." Consequently, the activation of a positive feedback loop will operate to the detriment of the individual. This is depicted in Fig. 9.3. As Scriver (1984) has pointed out, this concept of dishomeostasis is relevant not just

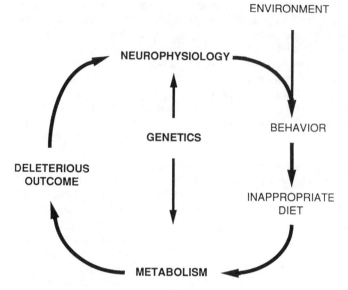

Fig. 9.3. Diagram of a positive feedback loop with deleterious effects. The paradigm illustrated is appropriate for complex situations like alcohol abuse, where genetic differences in both metabolism and neurophysiology can exacerbate the deleterious effects of alcohol.

to Mendelian defects of metabolism, but also to disease in general, a point first emphasized by Garrod (1931). While many genetic disorders of metabolism will influence positive feedback loops, favism being just one example, there are also numerous instances where neurophysiologic pathways also play a role, as indicated in Fig. 9.3. The case of alcohol abuse provides an illustration of the complex interactions that can underlie a positive feedback loop.

Unlike the situation in mice, where propensity for increased alcohol intake appears mediated by a single Mendelian locus (Goldman et al. 1987), the genetic underpinnings of excessive alcohol consumption in our species are much more complex. Cloninger et al. (1988) have recently shown that the behavioral reaction to environmental stimuli results from the interaction between three independently heritable personality axes, each modulated by one of the three major monoaminergic pathways (serotonin, dopamine, norepinephrine). As a consequence, certain individuals whose specific personality profile is due to the joint action of the three monoamine neuromodulation pathways, tend to develop the behavioral patterns that lead to alcohol abuse when placed in a specific high-risk sociocultural environment (Cloninger 1987). Once the pattern of alcohol abuse has been established, a positive feedback cycle tends to develop with consequent deviations from physiologic and behavioral norms. The magnitude of the deviation in terms of the degree of antisocial behavior appears to depend on the particular personality profile of the individual (Cloninger 1987).

In addition, as suggested by Fig. 9.3, underlying differences in the ability to metabolize alcohol may magnify the problem for some individuals. For example, it now seems likely that the probability of damage to the fetus of pregnant women who drink is a function of differences in metabolizing ethanol and polyamines

(Sessa et al. 1987). Different allelic forms of the mitochondrial aldehyde dehydrogenase enzyme may also result in different probabilities of alcoholic liver disease (Shibuya and Yoshida 1988). Such metabolic differences, which are probably genetic in origin, are also likely to influence the action of specific neuromodulation pathways which may lead to an exacerbation of behavioral profiles that result in alcohol abuse. Hence, alcohol abuse appears to arise from an extremely detrimental, genetically defined, positive feedback loop that once established is exacerbated by both metabolic and neurophysiologic pathways.

It will be noted that such detrimental positive feedback loops require an external perturbation. In the instance of alcohol abuse, it is fairly clear that sociocultural trends in societal use of alcohol have had a marked impact on the inheritance of alcohol abuse (Cloninger et al. 1988). Consequently, attempts to reduce the major social and medical complications of alcohol abuse are more likely to be effective if they concentrate on reducing the magnitude of the external stimuli, i.e., by lowering overall societal consumption, as opposed to developing programs targeted towards high-risk individuals (Sigvardsson et al. 1986).

More Complex Models: The Example of Obesity

When the consequences of the dietary–behavioral interaction are as complex as is the case with obesity, the number of possible interactions between genetic and environmental pathways also increases. Fig. 9.4 indicates how interaction between all four major categories of causal influence may result in a deleterious

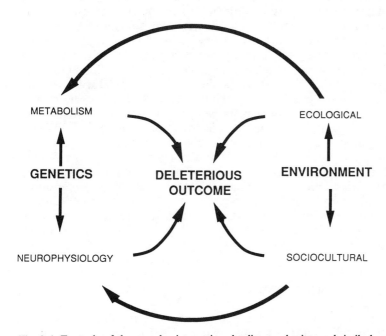

Fig. 9.4. Example of the complex interactions leading to obesity, and similarly complex outcomes. Note that all four causal pathways (metabolic, neurophysiologic, ecologic and sociocultural) are involved, each establishing its own positive feedback loop contributing to a deleterious effect over all.

outcome. As was suggested by the basic paradigm of Fig. 9.1, neurophysiologic pathways tend to interact with sociocultural pathways, while metabolic pathways tend to interact with components of the ecologic environment. In the case of obesity, the influence of both types of environmental pathway on obesity is well known, while the influence of genetically defined differences in metabolism has also been defined.

The role played by the availability of high caloric foods (i.e. the ecologic environment) on the distribution of obesity is well documented. These ecologic factors represent a major cause of the differences in the prevalence of obesity between populations. As a consequence, in analogous fashion to the argument for constraining alcohol abuse, it has been suggested that changing dietary profiles at the societal level represents the most effective strategy for diminishing the morbidity due to obesity (Blackburn 1984). However, as the studies of Worsley et al. (1984), Takahashi (1986) and others indicate, the sociocultural perception of body form also plays an important role in influencing food preference and hence food intake. Hence, both kinds of environmental factors need to be considered.

With respect to the influence of genetics, there is now little doubt that genetic factors play an important role in determining the development of obesity (Mueller 1983; Stunkard et al. 1986). However, the influence of the two main causal pathways is not as clearly demarcated as is the case for the environmental variables. While the influences of neurophysiologic factors have been defined in terms of appetite control, the genetic basis for these differences is not well established. On the other hand, the influence of genetically mediated differences in overall metabolic rate is better established. By evaluating total resting metabolic rate in terms of energy expenditure it has been possible to test explicitly the hypothesis implicit in Herrick's rhyme, namely that individuals with a low metabolic "set point" are more likely to become obese (Fox 1973). Two related studies have shown that low levels of resting metabolic rates not only represent a significant risk factor for increased weight gain (Ravussin et al. 1988), but that low levels of energy expenditure also exhibit familial transmission (Bogardus et al. 1986). Moreover, it now appears that the influence of low metabolic rate in mothers results in an increased tendency for weight gain in their children within the first year of life, suggesting genetic transmission (Roberts et al. 1988). While the genetic underpinnings of these differences in metabolic rates are likely to be complex, involving a multitude of loci, these results suggest a genetically defined causal mechanism for the familial aggregation of obesity. With respect to the relative importance of genetic vs. environmental factors, it is tempting to propose that within a community, familial aggregation of obesity is largely due to genetically mediated differences (such as intrinsic metabolic rates), while the differential distribution of obesity between population is likely to be due to the joint action of ecologic factors and sociocultural perceptions.

Consequences of Feedback Loops

In general, the existence of feedback loops has a major determining factor on the consequences of specific dietary or behavioral profiles. In some instances, as discussed above, the feedback involves the entire dietary–behavioral interaction axis, whereas in many others only a single component is involved. As noted above, genetic factors influence a variety of "detrimental" positive feedback loops that depend on a dietary intake for the initial perturbation. The impact of mutations at the low density lipoprotein (LDL) receptor on circulating levels of cholesterol represents a striking example (Brown and Goldstein 1986). Mutations within the LDL receptor gene, on chromosome 19, lead to diminished binding of LDL particles resulting in unabated cholesterol synthesis and greatly increased levels of LDL cholesterol. The mutation causes an interruption of the negative feedback loop normally required for optimal physiology. Such genetically compromised individuals have extremely elevated cholesterol levels and a substantial risk of an early heart attack (before age 40). In some populations, a single mutation at the LDL receptor locus appears to be responsible for a substantial portion of cases of early heart attacks in the community (Hobbs et al. 1987), whereas in others several mutations are involved. However, a study of putative gene carriers in Utah suggests the morbid consequences of this gene are not inexorable (Williams et al. 1985). If the environment eliminates such high-risk components as dietary fat, smoking, etc. a normal life span appears to be possible. Thus the dietary–behavioral consequences of feedback loops depend on the interaction between genotype and environment.

In this context, the recognition that genotype–environmental interactions underlie the origin of positive feedback loops is an elaboration of the concept of "dishomeostasis" (Scriver 1984), since the detrimental outcome results from the exposure of a susceptible genotype to a deleterious environment. The underlying genetic etiology may be as simple as a single locus, or as complex as a set of interacting major genes plus polygenic background. Overall, there appears to be an inverse relationship between the simplicity of genetic etiology and prevalence of dishomeostasis, single locus traits (e.g. phenylketonuria) being rare while more complex traits (obesity, alcohol abuse) tend to be common.

In either case, appropriate modification of the environment can establish a negative feedback loop with beneficial results. In the case of single locus traits such as phenylketonuria (PKU), dietary modification can be highly selective and targeted to specific mutations (Gutter et al. 1987; DiLella et al. 1988). For more complex problems, genotypic susceptibility is likely to reflect the selection for homeostasis in scarce environments (Neel 1977) and will be both ubiquitous and multifaceted. As a consequence, the strategy of mass intervention (Blackburn 1984; Sigvardsson et al. 1986) is most likely to be successful in eliminating positive feedback loops. Hence the goal of identifying the genetic components of dietary–behavioral interaction will not only define etiology but will also allow a more focused approach to environmental modifications.

Measurement Issues: Defining Intermediate Phenotypes

The complex nature of dietary–behavioral interactions implies that attempts to identify genetic components will require the precise measurement of a wide array of attributes. As suggested by Fig. 9.1, irrespective of whether the causal pathway is metabolic or neurophysiologic, adequate measurements of both extrinsic variables and intrinsic variables will be required if the genetic analysis is to have validity. In terms of the paradigm of causality, intrinsic attributes should be defined in terms of the gene products that influence the interaction. To the extent that genes, and their products, cannot be directly observed, measurement of intrinsic attributes will require defining an "intermediate phenotype" (Sing et al. 1986), in terms of a biochemical or physiologic marker of gene expression. Alternatively, when the metabolic pathways are well defined, as is the case for lipid metabolism, variability in candidate genes can be directly measured (Luisis 1988).

Given the complexity of dietary–behavioral interactions and the diverse factors which may be involved, the initial specification of the causal paradigm will dictate the types of variables that will require measurement. Table 9.1 outlines some of the major categories of variables that will need to be carefully analyzed, the particular selection being dependent on the causal pathway being evaluated. The use of a specific assay to measure an intrinsic attribute will define the intermediate phenotype for that study. If dietary–behavioral interactions are the end point of interest, considerable attention will have to be paid to identifying critical environmental factors. In general, the kinds of environmental variables listed in Table 9.1 are also "intermediate" in the sense that they are not likely to be causally implicated in the interaction, but rather act as surrogate measures for specific causal variables.

Table 9.1. Examples of the kinds of attributes that need to be measured, along with the type of technique that is required for accurate quantitative definition of the trait

Attributes	Assays
Intrinsic	
Metabolic	Enzymatic, immunologic, molecular
Physiologic	Biochemical, in vivo perturbation (e.g., GTT)
Morphologic	Physical measurement, imaging (e.g., ultrasound)
Neurophysiologic	Biochemical (e.g. monoamine assay), psychometric
Extrinsic	
Ecologic: sociodemographic (e.g., income, education)	Individuals: questionnaires Populations: census trait data
Ecologic: geographic	Questionnaires
Sociocultural	Questionnaires to identify cultural norms; familial norms

The relationship between an intermediate phenotype and more specific attributes of genetic causation is illustrated by the phenomenon of lactose intolerance. Since the fundamental problem is due to inadequate concentrations of lactase in the brush border of intestinal eithelial cells, measurement of lactase activity is the variable of choice. However, biopsy of the small intestinal mucosa to measure lactase directly is impractical for epidemiologic studies. Consequently, indirect methods, such as the hydrogen breath test following a lactose load, or the rise in blood glucose following a lactose load, represent a practical solution and an appropriate intermediate phenotype. Analysis of this intermediate phenotype revealed the single recessive inheritance of the trait (Lisker et al. 1975). Without an understanding of the underlying metabolic processes and the consequent development of an appropriate measurement, the phenotype of interest – milk intolerance due to lactose deficiency – could not have been studied. Similarly, identification of lactase as the key enzyme in lactose intolerance identified the candidate gene and led to the definition of the genetic basis for the metabolic effect in terms of a single biallelic locus, with each of the three genotypes having distinct levels of enzyme activity (Ho et al. 1982).

The definition of an intermediate phenotype is also required by studies that aim to test hypotheses about the influence of neurophysiologic, or behavioral, pathways. In order to define the genetic basis of the monoaminergically modulated personality components, Cloninger et al. (1986) had to study the distribution of personality profiles amongst adopted and natural children. Not until the probable role of genetic factors and environmental factors in establishing somatization patterns had been deduced, was it possible to investigate the interaction between these behavioral profiles and patterns of alcohol abuse. In both cases, metabolic and neurophysiologic, the formulation of a conceptual model and the subsequent identification of the intermediate phenotype was a prerequisite for discriminating between competing hypotheses (Gilligan et al. 1987). The principle of identifying the appropriate set of measurements that define the relevant intermediate phenotype is an important component of successful studies in genetic epidemiology (Sing et al. 1986).

Study Design and Analytic Strategies

The development of an adequate study design requires obtaining observations to facilitate the contrast between competing hypotheses leading to unambiguous conclusions. In a laboratory setting, the hallmark of a good study design is the establishment of an experimental contrast that directly tests the hypothesis of interest. Although direct experimental control is often impossible in population-based studies, the necessity of developing a contrast between competing hypotheses is equally important. Lacking experimental control, the development of an informative study design and the incorporation of appropriate analytic strategies becomes mandatory if genetic hypotheses are to be tested.

In the case of laboratory studies, once an informative study design has been defined, the primary emphasis centers on the procedures required to obtain unambiguous measurements, with a reduced emphasis on the statistical analysis of the observations. By contrast, in population-based, epidemiologic studies, where experimental manipulation is not possible, the development and use of the statistical analysis becomes an integral component of good study design. In the case of genetic epidemiology, where direct measurement of genetic effects is often impossible, clear-cut hypothesis testing requires the marriage of an appropriate analytic strategy to an informative study design. As a consequence, the field of genetic epidemiology tends to be dominated by an emphasis on statistical procedures, each procedure corresponding to a specific study design (Elston and Sobel 1979). Besides the necessity of obtaining adequate measurements of an "intermediate phenotype" (see above), a successful genetic epidemiological study also requires congruence of study design and analytic procedures.

Within the field of genetic epidemiology four broad categories of study design can be recognized, each category leading to different levels of genetic inference.

Table 9.2 reviews these categories in terms of the relationship between the type of contrast that can be achieved within each category and the underlying conceptual model that can be evaluated. Examples of study designs that are commonly used within each category are also given. In considering these distinct categories of study design, it will be noted that there is a gradient with respect to the quality of genetic inference that can be garnered. Exploratory studies tend to be primarily epidemiologic in nature and hence are limited in the amount of information they can shed on genetic causation. By contrast, study designs that incorporate an explicit test of genetic causation yield strong genetic inference, but only weak inference with respect to environmental factors. Recognizing the existence of this "inferential gradient" implies that study designs appropriate for one specific hypothesis may be wholly unsuited for another. Thus there is an implicit relationship between a given study design, its accompanying analytic strategy, and the underlying conceptual model.

In general, an investigation of the causal pathways underlying a dietary-behavioral interaction will require that the contrasts defined by the analytic strategy will embody some aspect of both genetic and environmental components of presumed causation. For environmental variables, the appropriate contrast can usually be achieved by standard epidemiologic strategies which endeavor to generate an operational distinction between "exposed" and "non-exposed." As in standard epidemiologic studies, the lack of experimental control in population-based study designs can be overcome by sampling and statistical strategies that ensure an appropriate contrast. These are best exemplified by Rhoads' view (Chapter 7). However, the establishment of an ambiguous genetic contrast tends to be substantially more difficult.

Fundamentally, a genetic contrast requires that the outcome for an individual of a designated genotype be compared with the outcome for an individual possessing a measurably different genotype. In the largely experimental situations that obtain for plants and animals, such a contrast between identifiable genotypes can be achieved by means of carefully designed mating programs. For both practical and ethical reasons, such a strategy cannot be contemplated for man. Hence, for the majority of study designs the establishment of a genotypic contrast between individuals has to be achieved through indirect, rather than direct means. The two important exceptions to this generalization – association studies and linkage studies – lie at opposite poles of the inferential gradient with respect to genetic causation (Table 9.2). In each case, Mendelian markers are used to establish measurable genotypic contrasts between individuals, and in neither case are the Mendelian markers necessarily involved in the causal pathway.

Aside from the use of genetic markers (see below) the most commonly used strategy for establishing genetic contrasts is via Malecot's (1984) "coefficient of kinship" which defines the probability of sharing alleles identical by descent for a pair of individuals. Whether explicitly defined, or implicitly incorporated into the study design, genetic contrasts established by this coefficient are indirect. Even in simple designs, such as nuclear families, a single coefficient of kinship may mask underlying heterogeneity in the degree of genetic similarity between individuals. Thus siblings may share two, one or no alleles at a locus despite their equal coefficients of kinship. Accordingly, the successful use of indirect means to establish genetic contrasts (such as the coefficient of kinship) requires elaboration of the underlying genetic model by specifically defining each contrast on the

Table 9.2. Fundamental categories of study design in genetic epidemiology

Category	Conceptual model	Contrasts	Examples
Exploratory	Single implicit contrasts	Genetic and family environment	Familial aggregation
		Specific locus effects	Association studies
Quasi-experimental	Single explicit contrasts	Environmental only	Migrant studies
		Genetic only	Twin studies
		Genetic vs. environmental	Adoption studies
Fixed family clusters	Complex contrasts Explicit environment Implicit genetic	Measured environment Unmeasured genetic	Path analysis and family set studies
Testing explicit hypotheses	Complex genetic models – some environmental factors	Implicit genetic	Complex segregation analysis
		Explicit genetic	Linkage analysis
		Explicit genetic plus explicit environment	Regressive analysis

genetic axis in terms of the genotypic distribution defined by Mendelian segregation. This principle holds irrespective of whether the genetic component of causation is deemed to be a single locus, a series of major loci, or a set of polygenes. Hence, the distribution of genotypes must be definable for any specific constellation of genetically related individuals included in the study design (see Table 9.2). This principle has led to the series of analytic techniques for critically evaluating the aggregation of observations in related individuals. These techniques, which form the "hallmark" of genetic epidemiology (Morton et al. 1983), are primarily concerned with three statistical issues: establishing an appropriate design contrast; incorporating the consequences of ascertainment (Elston and Sobel 1979); and obtaining a valid test of hypothesis within the parameter space defined by the model of causation.

The direct strategy of utilizing polymorphic genetic markers to define measurable genotypic contrasts between individuals has seen little use in the past because of the scarcity of informative genetic markers. However, the recent advances in molecular biology have resulted in an explosion in the number of genetic polymorphisms to define genetic contrasts. The existence of thousands of genetic polymorphisms at the DNA level, plus the fact that many of these markers have been mapped into well-defined linkage groups (Donis-Keller et al. 1987), represents an immense resource. In particular, it is now possible to select genetic markers that provide specific genotypic contrasts involving candidate genes (Luisis 1988). This technique has already been used to identify possible relationships between apolipoprotein genes (Ordovas et al. 1986; Hegele et al. 1986; Templeton et al. 1988) and lipid profiles. The further development of highly polymorphic VNTR loci (Nakamura et al. 1988) increases the likelihood that a given candidate gene will be extremely informative.

Another important aspect of these mapped genetic markers is that they allow an assessment of the influence of discrete genotypes on segregating traits. As Boerwinkle et al. (1986) have shown, the use of identifiable genetic markers to discriminate between genetic backgrounds results in a substantial gain of statistical information. This principle has been used to define the influence of the apolipoprotein E locus on cholesterol levels (Sing and Davignon 1985; Boerwinkle et al. 1987). The demonstration that a single, identifiable Mendelian locus can have a major influence on the quantitative variability at the phenotypic level, indicates that the general application of this strategy, using mapped DNA polymorphisms, will be an efficient way of identifying the genetic basis of phenotypic variability (Weller 1986).

Exploratory Analyses

As suggested by Table 9.2, the first step in identifying the causal pathways of a dietary–behavior interaction will require the application of descriptive analyses that are essentially exploratory in nature. While such analyses do not permit extensive discrimination between competing hypotheses, they can indicate whether genetic causation is likely. If so, more elaborate study designs can then be set up to evaluate the specific genetic mechanism. Table 9.3 reviews some aspects of study design that apply to the two main exploratory strategies in genetic epidemiology. The main distinction between these different strategies is whether the genetic effects can be directly measured or not.

Table 9.3. Exploratory studies to identify genetic determinants: some issues of study design

---

Determining association with genetic markers: measured genetic contrasts
Identification of "candidate" genes
Case–Control study design
  Match on epidemiologic attributes?
Statistical methodology
  Contingency tables vs. logistic regression
Pitfalls in interpretation
  Influence of genetic and/or environmental heterogeneity
  Causation vs. association

Determining extent of familial aggregation: indirect genetic contrasts
Selection of probands
  Random vs. specific ascertainment criteria
Identification of relatives
  Fixed cluster vs. sequential sampling
Statistical methodology
  Problem of correlated observations
  Adjustment of rates by appropriate concomitant
Pitfalls in interpretation
  Correlated genes vs. correlated environments
  Pervasiveness of ascertainment bias

Association with Genetic Markers: A Direct Strategy

One exploratory approach to determine whether genetic factors may influence the outcome of a particular dietary–behavioral interaction, is to determine whether the outcome is associated with Mendelian genes. Provided the right genes are selected, this has been a productive approach as indicated by the plethora of studies showing association between human lymphocyte (histocompatibility) antigens (HLA) and autoimmune disease. The logic behind such studies is simple: if a particular outcome shows a statistically significant association with a well-defined Mendelian segregation then genetic involvement is likely. The application of this concept to a well-defined study design is perhaps less straightforward (George and Elston 1987), but if couched in epidemiologic terms certainly feasible. Since such studies are representative of the generic case–control study design, (reviewed by Rhoads in Chapter 7) they have all the advantages of case–control studies, as well as the pitfalls.

The first issue to be resolved when applying this preliminary strategy to investigate the interaction between diet and behavior, is to identify an appropriate "candidate" gene (or genomic region). The analyses of HLA associations have been so successful in part because the majority of diseases studied are those in which immunologic disorders are implicated, and in part because the extensive variability of the HLA system makes it extremely informative. Hence, it will be necessary to build on existing biochemical and metabolic knowledge to identify potentially informative candidate genetic markers that should be studied. For example, in the case of alcoholism both alcohol dehydrogenase and aldehyde dehydrogenase are involved in oxidative metabolism of ethanol and there is preliminary evidence that isoenzymes at both loci may influence the metabolism of alcohol, and hence impinge on the risk of alcoholism (Bosron and Li 1986; Shibuya and Yoshida 1988). DNA polymorphisms for these genes will thus prove informative. For other metabolic aspects of dietary behavior, such as lipid profiles, a variety of candidate genes exist (Luisis 1988). Association studies using polymorphisms at apolipoprotein genes have identified significant genetic components not only of lipid variability (Sing and Davignon 1985; Hegele et al. 1986; Ordovas et al. 1986; Templeton et al. 1988) but also of obesity as well (Rajput-Williams et al. 1988).

An equally important issue centers around the problem of adequate study design, in particular whether cases and controls have been appropriately matched for potentially confounding variables. For genetic studies, this necessarily involved matching on genetic background, as well as on pertinent sociodemographic variables. Since the distribution of genotypes defines the distribution of key "exposure" variables, cases and controls should have equivalent probability distributions of genetic background. Unfortunately this important requirement is often overlooked, which places the interpretation of results from such studies in jeopardy. Even if efforts are made to identify, and standardize for different ethnic background, this can prove a complex task, as indicated by the recent study of the relationship between apolipoprotein genes and heart disease (Ordovas et al. 1986). Even when environmental variables and genetic background have been standardized (or matched), statistical methods are usually required to incorporate these variables explicitly. For this reason, logistic regression (or its equivalent) is to be preferred over simple contingency analysis.

Interpretation of the results from such studies should be cautious. First it is often difficult to completely eradicate some of the potential problems caused by inadequate matching of cases and controls. Hence doubt remains whether the association is due to the involvement of the genomic region in which the marker is located as opposed to different allele frequency distributions. Even if the study design was impeccable, problems of interpretation still remain since a variety of genetic effects can give rise to allelic association. The most common interpretation of a positive association is that the genetic marker itself is etiologically involved in the outcome, or is in close physical proximity to a causally involved genetic locus. However, population structure can give rise to allelic association even when the loci are not physically close. In addition, epistatic interactions can occur between the marker locus and another that is causally involved. Thus a significant association resulting from a carefully designed case–control study should be the impetus for a follow-up study designed to test critically the hypothesis of genetic involvement, and not a conclusive result.

Familial Aggregation: Indirect Strategy

An alternative preliminary exploratory analysis is to define the extent of familial aggregation of the consequences of dietary–behavioral interactions, since substantial genetic etiology will result in significant familial clustering of the trait. Despite the apparent simplicity of this strategy, as indicated in Table 9.3, there are a number of aspects of study design that require attention. Ascertainment of "probands" and their biological relatives should be carried out in an epidemiologically defensible manner. This necessarily requires the *a priori* definition of a causal paradigm in order to identify the ecologic and sociocultural correlates in which the sampling scheme will be based. With respect to sampling relatives, there is considerable advantage in evaluating the phenotypic distribution in a set of relatives that is uniformly defined for all probands. This is tantamount to establishing a "fixed cluster" sampling framework, which has the added advantage of allowing more specific tests of hypothesis (Chakraborty and Schull 1979).

The two major problems relate to statistical methodology and interpretation of results. Since the aggregate set of observations for a fixed cluster not only lacks independence, but also has differing distributions of covariates within the cluster, simple correlations and relative frequencies are error prone. The use of linear models that incorporate temporal covariates is recommended, though these are infrequently used. Interpretation of statistically significant familial clustering can also be problematic since the influence of shared environments can be just as powerful as genetic factors in causing aggregation. Thus, for lipoprotein profiles, which are known to have a strong genetic component (Rao et al. 1979; Iselius 1988), a considerable proportion of the familial aggregation can be attributable to the shared "household environment" (Hasstedt et al. 1985). Similarly the "cultural transmission" (Cavalli-Sforza and Feldman 1981) of behavioral profiles and attitudes can lead to familial aggregation of dietary outcomes that extend far beyond the immediate nuclear family. Hence, for a situation where behavioral and underlying metabolic factors interact with a distribution of dietary profiles, the demonstration of familial aggregation can be difficult to interpret without ambiguity. Nevertheless, the demonstration of familial clustering has been a productive strategy for identifying the probable influence of genetic factors in traits as complex as alcoholism (Cloninger 1987; Cloninger et al. 1988) and obesity (Hartz et al. 1977; Mueller 1983). The demonstration of familial aggregation for intermediate phenotypes such as resting metabolic rate (Bogardus et al. 1986; Ravussin et al. 1988) is also important suggestive evidence for the role of genetic factors in the complex phenotype of obesity.

Adoption Studies and Twin Studies: Studies Involving a Single Contrast

The second group of research strategies listed in Table 9.2 are "quasi-experimental" in the sense that the study design explicitly defines a single contrast for one of the prospective causal pathways – either genetic or environmental. In order to constrain the experimental contrast to a single axis, it is necessary to eliminate, or at least minimize, variability in the alternative causal path. There are two relatively common situations where this occurs: twins, where the environment is presumed constant and the "experimental contrast" is genetic; and adoptions. where the reverse is usually the case. However, as indicated by the comparison in Table 9.4, these two types of study design are not entirely complementary. Adoption studies have the greatest potential for establishing well-defined contrasts with internal controls, while twin studies tend to be much easier to carry out.

Table 9.4. Quasi-experimental studies: Contrast between adoption studies and twin studies

Experimental design	Adoption studies	Twin studies
Genes constant, environmental contrast	Yes	No
Environments constant, genetic contrast	Yes	Yes
Environmental and genetic contrasts	Yes	No
Simplicity of design	Yes	Yes
Ascertainment of sample units	Difficult	Easy
Potential for ascertainment bias	High	High
Implicit assumptions	Few	Many
Ability to test specific genetic model	Low	Very low
Ability to test specific environmental model	Moderate	Very low

For either study, the issue of ascertainment is of paramount importance. While biased ascertainment appears to be frequent in twin studies, it is a potential problem for adoption studies. Although much easier to accomplish, twin studies suffer from the inherent disadvantage that the presumption of environmental similarity is almost never met. Hence the contrast in concordance rates for monozygous twins compared with dizygous twins will be confounded by the correlation between genetic similarity and environmental similarity. The fact that twins do not necessarily represent the population at large, plus their relative scarcity and the necessity of stratifying by sex, are additional obstacles. Nevertheless, despite these problems, studies of twins have been an informative strategy for identifying the genetic influence on such complex traits as resting metabolic rates (Fontaine et al. 1985).

Adoption studies offer more potential since they allow the establishment of design contrasts that include internal controls. Thus, if adopted children are ascertained, the contrasting correlations with natural and adoptive parents living in different households can be standardized by the parent–offspring correlations derived from within households. Although not often attempted, selection of children as "cases" could be contrasted with selection of parents as "cases", which stratification by environmental risk factors would also be informative. The power of adoption studies is illustrated by their success in defining the genetic basis for alcohol abuse, as well as for the associated personality profiles (Cloninger 1987; Cloninger et al. 1986, 1988). Adoption studies have also proved extremely informative in defining the probable nature of genetic effects in obesity. In two separate studies – one in Scandinavia (Stunkard et al. 1986) and one in middle America (Price et al. 1987) – the distribution of obesity in children was shown to be a function of the status of their natural parents, rather than their adoptive parents. An important advantage of adoption studies is their straightforward study design and simple statistical analysis. The major disadvantage is the difficulty in obtaining detailed information about natural parents once a child has been adopted. As a consequence it is generally difficult to perform such studies apart from in Scandinavian countries.

Lastly, there is one important, but frequently overlooked, aspect of monozygotic twin studies. Since monozygous twins share identical genotypes any difference in phenotype must be attributable to environmental differences. Thus, under the appropriate conditions, monozygous twins can be considered in terms of a case–control study design in which the genotype is held constant, and only environmental factors can give rise to any discordance between twins (Ward 1983). This kind of study design appears to have considerable utility for investigating the presumed genetic contribution to such complex interactions as neuropeptide profiles and personality, as indicated by the recent analysis of the effect of nutritional factors on steroid hormone levels (Bishop et al. 1988). The use of this strategy provided preliminary evidence that dietary intake influences the plasma sex steroid levels in males, with significant associations between specific nutrients as well as total caloric intake and the hormones testosterone, follicle-stimulating hormone, androstenediol glucuronide and testosterone glucuronide. While overmatching is a potential problem for this study design, monozygous twins tending to be exposed to very similar environments , this only affects statistical power and estimates of attributable risk, rather than the basic hypothesis testing strategy. Similar strategies can be designed for certain fixed cluster sample units (Schull et al. 1970; Chakraborty and Schull 1979). For example, sibling sets, or even cousin sets, could be stratified on the basis of the degree of genetic homology at measurable genetic loci in an effort to identify environmental interactions. While such studies do not have the simple elegance of monozygous twin comparisons, they could be more practical in efforts to identify the relative impact of deleterious environments on susceptible genotypes in the general population.

Fixed Clusters: Partitioning Genetic and Environmental Effects

While samples of fixed clusters of relatives can be used to define the familial aggregation of a trait, in an exploratory sense, this sampling strategy also leads to the third category of research strategies. As indicated by Table 9.2, the aim of this research strategy is to partition out the relative contribution of genetic vs. environmental effects. The statistical underpinnings of this strategy hark back to the classical partitioning of variance components by plant and animal breeders. In employing this strategy, the underlying model of genetic causation is deemed to be essentially polygenic, with only minimal influence of identifiable single genes. Despite this constraint, the value of this analytic strategy is that it permits an intensive evaluation of the influence of sociocultural and ecologic effects. By appropriate choice of fixed clusters, the strategy facilitates the estimation of the influence of hierarchically defined environmental factors such as the characteristic environments of an ethnic group; of a socioeconomic group; or of a single household. More importantly, by appropriate modelling and sample design, this strategy permits hypothesis testing about the familial transmission of sociocultural factors, and hence tests theories of cultural transmission (Cavalli-Sforza and Feldman 1981). Since the majority of dietary–behavioral interactions are likely to be influenced by a complex set of environmental factors, exploiting analytic strategies that utilize fixed clusters is likely to be an extremely informative approach.

Table 9.5 indicates that the framework of a fixed cluster design incorporates two main kinds of analytic strategy. The distinction between the two approaches relates to the complexity of contrasts contained within a specified sampling unit, versus the complexity defined by the causal model being tested. In the strategy exemplified by path analysis, the sample units have simple structure (typically pairs of relatives, or simple nuclear families) with the consequence that the various design contrasts are conmingled within a single sample unit. However because the study design incorporates an entire set of independent sample units (e.g. sibships, parent–offspring pairs), the full set of contrasts can be defined in terms of a linear model leading to estimation of the underlying causal parameters.

Table 9.5. Fixed cluster samples

Objective	Partitioning of relative contribution of genetic transmission
	Cultural transmission of environmental components
Analytic strategies	Path analysis
	Regression analysis
Advantages	Epidemiologic sampling frame
	Genotype – environment interactions difficult to model

The alternative strategy, exemplified by the "family set" method (Schull et al. 1970), requires defining the sample unit in terms of a set of explicit contrasts and subsequently estimating the coefficients of causation by direct application of standard regression techniques. Thus, in the initial proposal for a family set, the sample unit includes relatives of the first degree (sibs), third degree (cousins) and genetically unrelated individuals (spouse and a random control). The distribution of these genetically related individuals within and between households allows explicit mean-square contrasts for estimating the relative influence of polygenic factors, plus the household environment, and the "neighborhood" environment. This concept was subsequently elaborated to a diverse set of fixed cluster sampling units, each embodying a distinct set of causal contrasts (Chakraborty and Schull 1979). Since this approach hinges on sampling the entire fixed cluster, within which the design contrasts occur, the confounding influence of ecologic variables is reduced. The major drawback to this innovative approach is the fact that large samples are required for stable estimates, which conflicts with the requirement that each fixed cluster should encompass a fairly extensive set of relatives and unrelated individuals. Since most communities will not yield more than a handful of such sampling units, this procedure is difficult to apply except in special circumstances.

Since path analytic techniques offer greater flexibility in evaluating the influence of complex causal pathways this alternative strategy is likely to have greater application in the field of dietary–behavior interactions. Originally developed by Wright (1921), path analysis allows estimation of specific paths of causation due to exogenous variables (directly observable) and latent variables (unobservable). The estimation principle is based on solving a set of simultaneous equations that specify the relationship between an observed covariance structure and the presumed underlying causal parameters. Widely used in a diverse number of fields, especially econometrics and the social sciences, this strategy has become a useful tool for genetic epidemiologists (Morton et al. 1983). The technique originally involved specifying the relationship between a set of correlations between sets of first degree relatives in terms of two distinct causal factors: polygenic factors (genetic heritability) and socioculturally transmitted factors (cultural heritability) (Rao et al. 1976). Subsequently, the models of causation have substantially increased in both realism and complexity (Fulker 1988). The increased statistical and computational complexity of the models has unfortunately placed the more realistic models beyond the reach of most investigators.

Since the application of simple path analysis models is likely to be an informative strategy for obtaining an initial assessment of the underlying causal factors in dietary–behavioral interactions, an overview of the technique is presented. Since the set of observed correlations needs to be defined in terms of underlying causal factors, the explicit formulation of a causal model represents the first step. Figure 9.5 represents a simple example for observations derived from measurements on the set of first degree relatives that constitute a nuclear family, with both genetic and sociocultural factors being causally implicated (Ward et al. 1980). Note that the causal model is consistent with the paradigm in Fig. 9.1. The diagram depicts causal paths by arrows, with squares representing the observed, measured variables, while the circles represent the underlying latent (unobservable), causal factors. Note that the influence of measurable, exogenous factors on the phenotype P, is represented by the symbol I, sometimes referred to as an "environmental index" (Rao et al. 1976). The subscripts M, F, D1, D2, S1, and S2 identify the specific individuals who make up a generalized nuclear family: mother, father, daughter(s) and son(s), while the causal influences are represented by G for genetic factors (assumed polygenic) and C for sociocultural factors. It should be stressed that path analysis can be exceedingly useful for other sample designs such as twins and adoption studies (Fulker 1988).

The causal model depicted by Fig. 9.5 indicates that while each offspring inherits his/her genetic determinants from each parent, irrespective of sex, there is provision for sex-specific transmission of sociocultural factors. Thus, there are four partial regression coefficients that identify cultural transmission from father to son, f_{FS}, father to daughter, f_{FD}, mother to son, f_{MS}, and mother to daughter, f_{MD}. As a consequence, the model allows for distinct categories of sociocultural correlations between spouses (u), brothers (v_S), sisters (v_D) and sibs of unlike sex (w). Lastly, the coefficients of determination that define the relative influence of

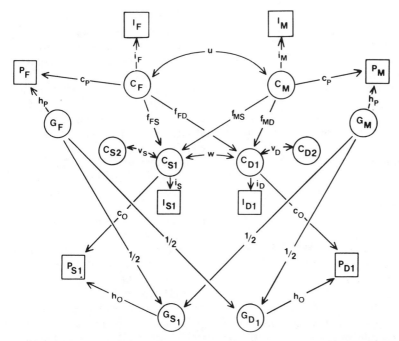

Fig. 9.5. Path diagram depicting the causal pathways that can exist in a nuclear family. P and I refer to the observable phenotype and environmental correlates, respectively, while G and C refer to the underlying genetic and sociocultural unobservable causes. The single-headed arrows define transmission of causal influence, while the double-headed arrows define correlations. Subscripts identify the specific family member. (See text for further details.)

genetic factors on the phenotype are defined by h_P (parents) or h_O (offspring), and for culturally transmitted factors by c_P and c_O respectively. The square of these coefficients of determination estimates the proportional contribution of genetic factors (h^2) and socioculturally transmitted factors (c^2) to the total phenotypic variance.

The path diagram can be used to relate the underlying causal determinants to the expected correlations by merely tracing the path from one observable attribute (denoted by a square) to another, via the specified paths of causation. Thus, the expected correlation between a father and son, $r(P_F, P_S)$ is given by:

$$h_P \cdot (1/2) \cdot h_O + c_P \cdot r_{FS} \cdot c_O$$

where $r_{FS} = f_{FS} + f_{MS} \cdot u$.

By following this principle in a consistent fashion, a set of equations is defined that relates all the observed correlations to the underlying causal factors assumed by the model. Solving these structural equations yields estimates of the magnitude of the various causal determinants. Explicit hypotheses of causation are tested by comparing the likelihoods obtained from contrasting sets of equations.

This informative strategy has been applied to a wide variety of situations that impinge on dietary–behavioral interactions. The analysis of lipid profiles is of particular interest, since path analysis has helped in defining the relative role of genetic and sociocultural factors on the distribution of various lipid constituents.

Using a model similar to Fig. 9.4, Rao et al. (1979) determined that genetic factors made a substantial contribution to the distribution of cholesterol and lipoprotein components in a Japanese-American population, whereas sociocultural factors had only a minor effect. A number of subsequent studies using path analysis have confirmed the importance of genetic heritability for the familial aggregation of total cholesterol, HDL-cholesterol and LDL-cholesterol, contrasted to the minimal genetic influence on triglyceride levels (Goldbourt and Neufeld 1986; Iselius 1988). Similarly, genetic heritability has been shown to exert a major influence on the distribution of intermediate phenotypes closer to the gene product, such as apolipoproteins A-I, B, and (to a lesser extent) apolipoprotein A-II (Hamsten et al. 1986). Hence despite the obvious contribution of major ecological differences to the overall distribution of lipid profiles (Blackburn 1984) the application of path analysis has served to emphasize the important role played by genetic factors.

Since path analysis is ideally suited for discriminating between the different causal pathways depicted in Figs. 9.1–9.3, it should be given more consideration as a strategy for analyzing dietary–behavioral interactions. The utility of path analysis for studying complex phenotypes is indicated by the successful linkage of genetic and environmental factors to such components of obesity as relative fat (Bouchard et al. 1988). Path analysis is also particularly valuable in assessing the influence of behavioral factors in which there is a substantial amount of sociocultural transmission within families (Rice et al. 1978; Fulker 1988). Hence, for many problems the estimation of the specific sociocultural transmission parameters will be more informative than a single global estimate of the overall influence of genetic (or environmental) factors (Ward et al. 1980). The ability to carry out multivariate analyses is also a substantial advantage (Vogler 1985).

Testing Explicit Hypotheses: Segregation, Linkage and Regressive Analyses

The final set of analytic strategies to be considered are those that emphasize explicit tests of hypotheses concerning the genetic etiology of a trait. Consequently, the conceptual model underlying these analytic strategies is firmly based on the principle of Mendelian segregation at one or several distinct loci. The essence of all these strategies is to evaluate the etiologic consequences of Mendelian segregation at loci whose alleles exert a measurable effect on the phenotype – either by confirming the existence of such loci or by mapping their physical position in the genome. Since the success of this strategy depends on detecting Mendelian segregation, care should be taken to define an "intermediate phenotype" that is relatively close to the primary gene product of the candidate gene. Wherever possible the intermediate phenotype should be defined in terms of the action of specific enzymes, or neurotransmitters, depending on which axis of causation is being investigated (Fig. 9.1). The appropriate definition of the intermediate phenotype is an important component in the successful testing of explicit genetic hypotheses, since the influence of Mendelian segregation is unlikely to be discerned in phenotypes that are far removed from primary gene action.

There are three main types of genetic analysis within this general category, each with a distinct set of sampling and analytic requirements (Table 9.6). Complex segregation analysis is primarily concerned with testing the hypothesis that Mendelian segregation at a single "major gene" exerts a significant influence on the phenotypic distribution. While the basic principles of complex segregation analysis for generalized pedigrees were established some considerable time ago (Elston and Stewart 1971), the successful application of this strategy required the development of algorithms for calculating conditional likelihoods in generalized pedigrees (Ott 1977, 1979). There are three fundamental parameters to be estimated in complex segregation analysis: the transmission probabilities, the degree of dominance, and the allele frequency. However, ancillary parameters, such as the ascertainment parameter and the form of the penetrance function, also need to be estimated as their values can impact on the ability to test hypotheses.

Table 9.6. A comparison of strategies for testing explicit Mendelian hypotheses

	Analytic strategy		
	Complex segregation	Linkage	Regressive
Genetic etiology			
Major locus	Yes	Yes	Yes
Polygenes	Yes	No	Yes
Multiple loci	Difficult	Yes	Yes
Sampling frame			
Extended pedigrees	Yes	Yes	Yes
Fixed clusters	Yes	Inefficient	Yes
Environmental aspects			
Penetrance	Yes	Yes	Yes
Cultural transmission	Yes	No	Yes
Shared environments	Difficult	No	Yes
Genotype–Environment	No	No	Yes

As originally construed, traditional segregation analysis had the drawback of being restricted to defining the consequences of Mendelian segregation at a single locus in the absence of polygenic factors. An important advance in overcoming this constraint was the development of the so-called mixed model in which each segregating genotype was associated with an underlying normal distribution, the mean of the distribution representing the genotypic effect and the variance in the joint influence of polygenic and environmental factors (Morton and MacLean 1974). The population distribution of the phenotype is thus deemed to be a mixture of underlying normal distributions, each corresponding to a Mendelian genotype. This technique initially proved even more difficult to implement in generalized human pedigrees and required an inordinate amount of computer time for all but trivially small pedigrees (Elston and Rao 1978; Ott 1979). However, with the development of "peeling" algorithms so that pedigrees could be transversed in an efficient manner (Cannings et al. 1978) plus an approximation technique for calculating likelihoods (Hasstedt 1982), this procedure became feasible for pedigrees that were both large and complex. The incorporation of the "mixed model" concept into the strategy of estimating transmission parameters led to the "unified" approach to complex segregation analysis (Lalouel et al. 1983). The fundamental parameters to be estimated are: the transmission parameters, the displacement between genotypic means, dominance relationships, and allele frequency; in addition the proportion of variance about the genotypic means that is attributable to polygenic factors should also be estimated.

Partly because of the difficulty in defining the appropriate intermediate phenotype, there are few examples where complex segregation analysis has successfully elucidated the genetic mechanism of a dietary–behavioral interaction. Thus far most attention has been focused on phenotypes such as lipid profiles, although one study had purported to find evidence for the recessive inheritance of fat-patterning (Hasstedt et al. 1989). Even in the case of lipid profiles, the evidence is somewhat inconsistent though the statistical evidence for influence of major genes on apolipoprotein B, and apolipoprotein A-I, seems secure (Iselius 1988; Luisis 1988).

Linkage analysis presumes that segregation of a major gene underlies the distribution of the phenotype and seeks to establish the extent to which the genetic determinants of the trait co-segregate with alleles of a known marker locus. Hence, the recombination fraction is now a fundamental parameter to be estimated, while the characteristic attributes of the etiologic major gene (dominance, allele frequency) become ancillary parameters, along with penetrance. In analogous fashion to segregation analysis, efficient application of linkage analysis in human pedigrees also awaited the development of computer techniques that allowed the accumulation of likelihoods across generalized pedigree structures (Ott 1977). More recently, this strategy has been expanded to allow the simultaneous estimation of recombination for an entire set of loci, thereby allowing specific placement of loci within a genomic region (Lathrop et al. 1984).

While segregation analysis and linkage analysis share many statistical features (Table 9.6), they embody rather different assumptions and hence have their own set of constraints. The use of regressive models has the potential to remove some of these constraints, as well as yielding sharper tests of hypothesis. This third technique applies multivariate regression techniques to estimate the underlying causal components that determine the correlations between relatives sampled from a single pedigree of arbitrary complexity. The correlations are defined in terms of genetic covariances that can be completely specified for any pedigree relationship in terms of both major genes and polygenic factors, plus "residual non-genetic" covariances which can be due to cultural transmission or sharing of environments. In this instance, the regression equation is written in terms of the conditional distribution of an offspring's trait on their parents' traits, generalized to transverse pedigrees of arbitrary structure (Bonney 1984, 1986).

The potential of the regressive model approach is that it allows evaluation of the influence of environmental factors in analogous fashion to the family set of procedures, but with some of the flexibility of path analysis. Thus the distribution of correlations between relatives can be influenced by their sex, age and position in the pedigree, as well as by relevant epidemiologic variables. The ability to specify directly such environmental parameters in the model removes many of the constraints associated with applying complex segregation analysis, or linkage analysis to complex genotype–environmental interactions. Lastly, the extension of the regressive model aproach to incorporate the principle of linkage analysis for an arbitrary number of loci (Bonney et al. 1988) opens the possibility of analyzing complex traits in terms of interactions between different loci. Application of ths procedure to analyze the distribution of apolipoprotein levels in a complex pedigree not only confirmed the existence of a major gene, but also gave a sharper test of the hypothesis than had been obtained by a prior complex segregation analysis (Bonney et al. 1988). Consequently, despite the requirement for more statistical development of this approach, the advent of the regressive model holds considerable promise for investigating the genetic underpinnings of the relationship between diet and behavior.

Prospectus

Despite the growing awareness that genetic factors are likely to influence many aspects of dietary–behavioral interactions, the contribution of genetic epidemiology to the field has thus far been slight. There seem to be two main reasons for this: insufficient communication between geneticists and nutritionists, and the complexity of the problem. As opportunities are created for the two disciplines to meet, there will be an increasing recognition of the contribution that each field can make. When there is a deliberate attempt to find common ground, a whole host of productive research questions can be established, as exemplified by the recent workshop on genetic factors in nutrition (Velazquez and Bourges 1984). Similarly, such meetings offer an opportunity to determine which research strategies in genetic epidemiology can be most profitably employed in a nutritional context (Ward 1984).

The complexity of the problem imposes a further set of constraints if the methods of genetic epidemiology are to be used to identify the causal pathways underlying dietary–behavioral interactions. As stressed in the first part of this chapter, the first requirement is to determine a reasonable model of causality so that specific hypotheses can be tested. While the paradigms illustrated by Figs. 9.1–9.4 are not exclusive, they indicate the kind of specific causal pathways that need to be evaluated. Identification of a likely causal pathway in terms of biological phenomena (e.g. metabolic, neurophysiologic) and then determining an informative "intermediate phenotype" thus becomes the crux.

The relationship between the "intermediate phenotype" and the degree of specificity of the genetic model results from the fact that the most specific genetic models (e.g. complex segregation analysis) require the detection of Mendelian segregation and hence an "intermediate phenotype" that is not too far removed from primary gene action. If the "intermediate phenotype" is quite distant from the primary gene product then exploratory analyses (e.g. association analysis) will be more productive. Thus, although the influence of segregation at the apolipoprotein E locus on cholesterol levels has been clearly defined by association analysis (Sing and Davignon 1985; Boerwinkle et al. 1987), application of complex segregation analysis to cholesterol levels has yet to provide unambiguous evidence in favor of a major gene. If the phenotype is as complex as obesity (Fig. 9.4), where developmental changes result in radically different distributions of familial correlations for each component of obesity (Kaplowitz et al. 1988), naive application of the last set of analytic strategies (Table 9.6) is not likely to be productive.

Thus while the principles of genetic epidemiology have much to offer in understanding the causes of dietary–behavioral interactions, it is the more descriptive, exploratory strategies that will be most useful. In this regard, the application of association analyses in terms of candidate genes defined at the DNA level, appears to hold great promise. When appropriate "intermediate phenotypes" can be defined, as is the case for lipid profiles, the use of the candidate gene strategy has proved to be extremely informative (Luisis 1988). Since candidate genes for monoaminergic modulation pathways are in the process of being defined, this approach can be applied to neurophysiologic pathways as well.

The successful identification of specific genetic factors that contribute to dietary–behavioral interactions leads to more effective strategies for eliminating deleterious positive feedback loops (Fig. 9.3). When the elements of genetic causation cannot be identified, mass intervention is the only reasonable strategy (Blackburn 1984). If specific genes can be determined, intervention can be targeted to the appropriate genotype, as is now the case for PKU (Gutter et al. 1987; DiLella et al. 1988). This is not only more efficient, but also elicits greater compliance. The contribution of genetic epidemiology has been to open up the possibility of targeted intervention strategies for such complex problems as lipid disorders, lactose intolerance, and alcohol abuse. The continued application of the principles of genetic epidemiology, in the context of a growing array of molecular markers, to the complex field of dietary–behavioral interactions will undoubtedly lead to a substantial number of similar success stories.

References

Abraham SF, Beaumont PFV (1982) How patients describe bulimia or binge eating. Psychol Med 12:625–635

Bishop DT, Meikle AW, Slattery ML et al (1988) The effect of nutritional factors on sex hormone levels in male twins. Genet Epidemiol 5:43–59

Blackburn H (1984) Determinants of individual and population blood lipoprotein levels. In: Velazquez A, Bourges H (eds) Genetic factors in nutrition. Academic Press, New York, pp 105–136

Boerwinkle E, Chakraborty R, Sing CF (1986) The use of measured genotype information in the analysis of quantitative phenotypes. I. Models and methods. Ann Hum Genet 50:181–194

Boerwinkle E, Visvikis S, Welsh D et al. (1987) The use of measured genotype information in Man. II. The role of apolipoprotein E polymorphism in determining levels, variability and covariability of cholesterol, betalipoprotein and triglycerides in a sample of unrelated individuals. Am J Med Genet 27:567–582

Bogardus C, Lillioja S, Ravussin E et al. (1986) Familial dependence of the resting metabolic rate. N Engl J Med 315:96–100

Bonney GE (1984) On the statistical determination of major gene mechanisms in continuous human traits: regressive models. Am J Med Genet 18:731–749

Bonney GE (1986) Regressive logistic models for familial disease and other binary traits. Biometrics 42:611–625

Bonney GE, Lathrop GM, Lalouel J-M (1988) Combined linkage and segregation analysis using regressive models. Am J Hum Genet 43:29–37

Bosron WH, Li IK (1986) Genetic polymorphisms of human liver alcohol and aldehyde dehydrogenases and their relationship to alcohol metabolism and alcoholism. Hepatology 6:502–510

Bouchard C, Perusse L, Leblanc C et al. (1988) Inheritance of the amount and distribution of human body fat. Int J Obes 12:205–215

Brown MS, Goldstein JL (1986) A receptor mediated pathway for cholesterol homeostasis. Science 232:34–47

Cannings C, Thompson EA, Skolnick MH (1978) Probability functions on complex pedigrees. Adv Appl Prob 10:26–61

Cavalli-Sforza LL, Feldman MW (1981) Cultural transmission and evolution: a quantitative approach. Princeton University Press, Princeton

Chakraborty R, Schull WJ (1979) Fixed cluster designs in human genetic studies: interpretations and usefulness. In: Sing CF, Skolnick MH (eds) Genetic analysis of common disease. Alan R. Liss, New York, pp 343–362

Cloninger CR (1987) Neurogenetic adaptive mechanisms in alcoholism. Science 236:410–416

Cloninger CR, von Knorring A-L, Sigvardsson S, Bohman M (1986) Symptom patterns and causes of somatization in men. II. Genetic and environmental independence from somatization in women. Genet Epidemiol 3:171–186

Cloninger CR, Reich T, Sigvardsson S et al. (1988) Effects of change in alcohol use between generations on inheritance of alcohol abuse. In: Rose RM, Barrett J (eds) Alcoholism: origins and outcomes. Raven Press, New York, pp 49–74

DiLella AG, Huang W-M, Woo SLC (1988) Screening for phenylketonuria mutations by DNA amplification with the polymerase chain reaction. Lancet i:497–499

Donis-Keller H, Green P, Helms C et al. (1987) A genetic linkage map of the human genome. Cell 51:319–337

Drewnowski A, Bellisle F, Aimex P, Remy B (1987) Taste and bulimia. Physiol Behav 41:621–626

Elston RC, Rao DC (1978) Statistical modeling and analysis in human genetics. Annu Rev Biophys Bioeng 7:253–286

Elston RC, Sobel E (1979) Sampling considerations in the gathering and analysis of pedigree data. Am J Hum Genet 31:62–69

Elston RC, Stewart J (1971) A general model for the genetic analysis of pedigree data. Hum Hered 21:523–542

Fontaine E, Savard R, Tremblay A et al. (1985) Resting metabolic rate in monozygotic and dizygotic twins. Acta Genet Med Gemellol 34:41–47

Forman MR, Trowbridge FL, Gentry EM et al. (1986) Overweight adults in the United States: the behavioral risk factor surveys. Am J Clin Nutr 44:410–416

Fox FW (1973) The enigma of obesity. Lancet ii:1487–1488

Fulker DW (1988) Genetic and cultural transmission in human behavior. In: Weir BS, Eisen EJ, Goodman MM, Namkoong G (eds) Proceedings of the second international conference on quantitative genetics. Sinauer; Sunderland, pp 318–340

Garrod AE (1931) The inborn factors in disease. Oxford University Press, Oxford

Gilligan SB, Reich T, Cloninger CR (1987) Etiological heterogeneity in alcoholism. Genet Epidemiol 4:395–414

George JT, Elston RC (1987) Testing the association between polymorphic markers and qualitative traits in pedigrees. Genet Epidemiol 4:193–200

Goldbourt U, Neufeld HM (1986) Genetic aspects of arteriosclerosis. Arteriosclerosis 6:357–377

Goldman D, Lister RG, Crabbe JC (1987) Mapping of a putative genetic locus determining ethanol intake in the mouse. Brain Res 420:220–226

Gutter FG, Ledley FD, Lidsky AS et al. (1987) Correlation between polymorphic DNA haplotypes at phenylalanine hydroxylase locus and clinical phenotypes of phenylktonuria. J Pediatr 110:68–71

Hamsten A, Iselius L, Dahlen G, de Faire U (1986) Genetic and cultural inheritance of serum lipids, low and high density lipoprotein cholesterol and serum apolipoproteins A-I, A-II and B. Arteriosclerosis 60:199–208

Hartz A, Giefer E, Rimm AA (1977) Relative importance of the effect of family environment and heredity in obesity. Ann Hum Genet 41:185–193

Hasstedt SJ (1982) A mixed model likelihood approximation for large pedigrees. Comput Biomed Res 15:295–307

Hasstedt SJ, Kuida H, Ash KO, Williams RR (1985) Effects of household sharing on high density lipoprotein and its subfractions. Genet Epidemiol 2:339–348

Hasstedt SJ, Ramirez ME, Kuida H, Williams RR (1989) Recessive inheritance of a relative fat pattern. Am J Hum Genet 45:917–925

Hegele RA, Huange L-S, Herbert PN et al. (1986) Apolipoprotein B-gene DNA polymorphisms associated with myocardial infarction. N Engl J Med 315:1509–1515

Ho MW, Povey S, Swallow D (1982) Lactase polymorphism in adult British natives: estimating allele frequencies by enzyme assays in autopsy samples. AmJ Hum Genet 34:650–657

Hobbs HH, Brown MS, Russel DW et al. (1987) Deletion in the gene for the low density lipoprotein receptor in a majority of French Canadians with familial hypercholesteremia. N Engl J Med 317:734–737

Iselius L (1988) Genetic epidemiology of common diseases in humans. In: Weir BS, Eisen EJ, Goodman MM, Namkoong G (eds) Proceedings of the second international conference on quantitative genetics. Sinauer; Sunderland, pp 341–352

Kaplowitz HJ, Wild KA, Mueller WH et al. (1988) Serial and parent-child changes in components of body fat distribution and fatness in children from the London growth study, ages two to eighteen years. Hum Biol 60:739–758

Lalouel J-M, Rao DC, Morton NE, Elston RC (1983) A unified model for complex segregation analysis of quantitative traits, Am J Hum Genet 35:816–826

Lathrop GM, Lalouel J-M, Julier C, Ott J (1984) Strategies for multilocus analysis in humans. Proc Natl Acad Sci USA 81:3443–3446

Lisker R, Gonzalez B, Daltabuit M (1975) Recessive inheritance of the adult type of intestinal lactase deficiency. Am J Hum Genet 27:662–664

Lisker R, Aguilar L, Zavala C (1978) Intestinal lactase deficiency and milk drinking capacity in the adult. Am J Clin Nutr 31:1499–1503

Lisker R, Aguilar L, Lares I, Cravioto J (1980) Double blind study of milk lactase intolerance in a group of rural and urban children. Am J Clin Nutr 33:1049–1053

Luisis AJ (1988) Genetic factors affecting blood lipoproteins: The candidate gene approach. J Lipid Res 29:397–429

McCracken R (1971) Lactase deficiency: an example of dietary evolution. Curr Anthropol 12:479–517

McMurray MP, Connor WE, Cerqueria MT (1982) Dietary cholesterol and the plasma lipids and lipoproteins in the Tarahumara Indians: a people habituated to a low cholesterol diet after weaning. Am J Clin Nutr 35:741–751

Malecot G (1984) Les mathématiques de l'hérédité. Masson, Paris

Morton NE, MacLean CJ (1974) Analysis of family resemblance. III. Complex segregation of quantitative traits. Am J Hum Genet 26:489–503

Morton NE, Rao DC, Lalouel J-M (1983) Methods in genetic epidemiology. Karger, Basel

Mueller WH (1983) The genetics of human fatness. Yearbook Phys Anthropol 26:215–230

Nakamura Y, Carlson M, Krapcho K et al. (1988) New approach for isolation of *VNTR* markers. Am J Hum Genet 43:854–859

Neel JV (1977) Health and disease in unacculturated Amerindian populations. CIBA Foundation Symposium 49:155–177

Ordovas JM, Schaefer EJ, Deeb S et al. (1986) Apolipoprotein A-I gene polymorphism in the 3' flanking region associated with premature coronary death and genetic high density lipoprotein deficiency. N Engl J Med 314:671–677

Ott J (1977) Counting methods (EM algorithm) in human pedigree analyses: linkage and segregation analysis. Ann Hum Genet 40:443–454

Ott J (1979) Maximum likelihood estimation by counting methods under polygenic and mixed models in human pedigrees. Am J Hum Genet 31:161–175

Price RA, Cadoret RJ, Stunkard AJ, Troughton E (1987) Genetic contributions to human fatness: an adoption study. Am J Psychiatr 144:1003–1008

Rajput-Williams J, Knott TJ, Wallis SC et al. (1988) Variation of apolipoprotein B gene is associated with obesity, high blood cholesterol levels and increased risk of coronary heart disease. Lancet ii:1442–1446

Rao DC, Morton NE, Gulbrandsen CL et al. (1979) Cultural and biological determinants of lipoprotein concentrations. Ann Hum Genet 42:467–477

Rao DC,Morton NE, Yee S (1976) Resolution of cultural and biological inheritance by path anlaysis. Am J Hum Genet 28:228–242

Ravussin E, Lillioja S, Knowler WC et al. (1988) Reduced rate of energy expenditure as a risk factor for body-weight gain. N Engl J Med 318:467–472

Rice J, Cloninger CR, Reich T (1978) Multifactorial inheritance with cultural transmission and assortative mating. I. Description and basic properties of the unitary models. Am J Hum Genet 30:618–643

Roberts SB, Savage J, Coward WA et al. (1988) Energy expenditure and intake in infants born to lean and overweight mothers. N Engl J Med 318:461–466

Schull WJ, Harbury E, Erfurt JC et al. (1970) A family set method for estimating hereditary and stress. J Chron Dis 23:82–95

Scriver CR (1984) The Canadian Rutherford lecture: an evolutionary view of disease in man. Proc Roy Soc Lond (Biol) 220:273–298

Sessa D, Desidevio MA, Pevin A (1987) Ethanol and polyamine metabolism in adult and fetal tissues: possible implication of fetal damage. Adv Alcohol Subst Abuse 6:73–85

Shibuya A, Yoshida A (1988) Genotypes of alcohol-metabolizing enzymes in Japanese with alcohol liver diseases: a strong association of the usual Caucasian-type aldehyde dehydrogenase gene (ALDH 2-1) with the disease. Am J Hum Genet 43:744–748

Sigvardsson S, Cloninger CR, Bohman M (1986) Prevention and treatment of alcohol abuse: uses and limitations of the high risk paradigm. Soc Biol 32:185–194

Simoons F (1978) The geographic hypothesis and lactose malabsorption: a weighing of the evidence. Digest Dis 23:963–980

Sing CF, Boerwinkle E, Turner SJ (1986) Genetics of primary hypertension. Clin Exp Hypertens A8:623–651

Sing CF, Davignon J (1985) Role of the apolipoprotein E polymorphism in determining normal plasma lipid and lipoprotein variation. Am J Hum Genet 37:268–285

Stunkard AJ, Sorensen TIA, Hanis C et al (1986) An adoption study of human obesity. N Engl J Med 314:193–198

Takahashi E (1986) Secular trend of female body shape in Japan. Hum Biol 58:293–301

Templeton AR, Sing CF, Kessling A, Humphries G (1988) A cladistic analysis of phenotype associations with haplotypes inferred from restriction endonuclease mapping. II. The analysis of natural populations. Genetics 120:1145–1154

Velazquez A, Bourges H (eds) (1984) Genetic factors in nutrition. Academic Press, New York

Vogler GP (1985) Multivariate path analysis of familial resemblance. Genet Epidemiol 2:35–54

Ward RH (1983) Genetic epidemiology as a tool for investigating rheumatic disease. J Rheumatol 10(S10):46–49

Ward RH (1984) Genetic epidemiology as a potential tool in nutritional research. In: Velazquez A, Bourges H (eds) Genetic factors in nutrition. Academic Press, New York, pp 37–53

Ward RH, Chin PG, Prior IAM (1980) Effect of migration on the familial aggregation of blood pressure. Hypertension 2:43–54

Weller JI (1986) Maximum likelihood techniques for the mapping and analysis of quantitative trait loci with the aid of genetic markers. Biometrics 42:627–640

Williams RR, Hasstedt SJ, Wilson DE et al. (1985) Evidence that men with familial hypercholesterolemia can avoid early coronary death: an analysis of 77 gene carriers in four Utah pedigrees. JAMA 255:219–224

Worsley A, Peters M, Worsley AJ et al. (1984) Australian 10-year olds' perception of food. III. The influence of obesity status. Int J Obes 8:327–341

Wright S (1921) Correlation and causation. J Agric Res 20:557–585

Discussion

A. E. Harper

The contributors to this section were asked to explore the potential for, and the problems involved in, applying epidemiologic methods in investigations of interactions between diet and behavior.

Epidemiology is a discipline that deals with the incidence, distribution and control of disease. A major objective of epidemiologic investigations is to assess, as Rhoads, Vietze, and Ward all explain, the degree of association between the incidence of a disease or disorder and characteristics of both the population and the environment in which it occurs. Epidemiology might be defined simply as the study of the ecology or etiology of disease.

In traditional epidemiologic studies, the disease entity, outcome variable, or observable end point, is usually readily defined. The number of variables, i.e., personal traits and environmental characteristics, that may be associated with susceptibility or resistance to a disease is, however, immense. Distinguishing between associations that are pertinent and those that are not, and identifying relationships among them that may confound the results, is an overwhelming task, as all the contributors point out. Nonetheless, with readily recognizable and measurable end points such as death, or clearly defined signs and symptoms, such as elevated blood pressure; and with methods for measuring potentially influencing variables, such as fat intake or salt intake, the strength of associations between incidence of the disease and measures of potentially influencing factors, such as diet, can be estimated. (This is granted to be within limits of error that might be considered distressing in the biological sciences, and devastating in the physical sciences.) Such associations do not provide evidence of cause–effect relationships, but they do provide clues as to appropriate directions in which to search for such evidence.

If the cause of the disease is known, as Ward discusses in relation to lactose intolerance, the epidemiologic approach is simplified and can be used effectively to elucidate the etiology of the disease. I am skeptical, however, that the development of dairying on a large scale occurred long enough ago, as some speakers suggested, to explain the geographic distribution of lactase deficiency as the result of evolutionary pressure within milk producing societies for positive selection of lactase. Substantial milk consumption, even by children over the age of 2 years, is a recent phenomenon. It is much more likely that it was the other way around. High milk consumption in the modern environment permitted expression and detection of the enzyme or its absence, and thereby permitted epidemiologists to establish the population distribution of the gene.

When the underlying causes of a disease have not been established, however, as is the case with most of the chronic and degenerative diseases that are of major interest at present, and when the number of, and interactions among, genetic factors that influence susceptibility to these diseases are just now being elucidated, interpretation of results of epidemiologic observations remains tentative and often speculative. We can talk only of risk factors, probabilities within a large population that a particular characteristic is associated with development of the disease, not of causes or predictions of outcome for an individual, but we are getting closer.

If the complexities of epidemiologic investigations of diet and chronic and degenerative diseases are great, those of epidemiologic investigations of diet and behavior are much greater. If the behavior under study is a serious abnormality that persists, it may be considered analogous to a disease. Investigations of behavioral aberrations or mental diseases are beset with essentially the same problems as are traditional epidemiologic studies of physical diseases. The outcome variable, e.g., impaired behavior or mental development, is, however, much more difficult to define and quantify, although Rodin described some ways of dealing with this problem. Also, the potential confounding factors include a complex of interactions among social, economic and cultural variables, as is discussed in the previous sections. These are still more difficult to define and quantify.

In the context of this volume, however, we are not dealing so much with behavioral changes that are analogous to physical diseases as we are with acute, short-term responses to diet or food ingestion. These may represent only shifts in "normal" behavior, such as in sleep patterns. In fact, the behavioral change under study may represent part of the response of a regulatory system that contributes to homeostasis when the diet, a component of the external environment, impinges on the internal environment. A change in food preference, for example, may contribute to maintenance of a balanced intake of major food components. Such responses are transitory; they may change as the result of physiologic or metabolic adaptations and, unless the dietary influence is extreme, they may not only be difficult to measure, but even to detect. In fact, investigations of acute responses to diet may be not so much studies of effects of diet on behavior as they are studies of behavior as a component of regulatory mechanisms that contribute to maintenance of standard, or at least tolerable, levels of blood and brain metabolites.

Serious, persistent, behavioral effects of diet, at least insofar as we currently know, occur only after prolonged ingestion of nutritionally inadequate diets, or as the result of impairment, usually genetic, in the metabolism of a nutrient, as Vietze discusses. Among these would be included chronic malnutrition, nutritional deficiency diseases such as goiter, pellagra, scurvy, and starvation. We might also include here studies of the effects of toxic substances, such as mercury or lead, in diets.

Behavioral effects of malnutrition in children have been studied extensively, particularly in Mexico, Guatemala and India (Agarwal et al. 1987), using epidemiologic methods. The types of measures used have included tests designed to assess cognitive function, social competence, and coordination. In most of these studies, dietary intake data were not collected. The studies were based on assessments of the degree of malnutrition among the subjects. Results of earlier studies of associations between dietary intake and physical development have provided the basis for using clinical and anthropometric measurements to classify children as to degree of malnutrition.

Examination of associations between the outcome of the test scores and degree of malnutrition has revealed consistently that malnourished children show behavioral and cognitive deficits. In studies of this type, dose–response relationships between the degree of behavioral deficit and the degree of malnutrition have been observed. If associations between behavioral performance and dietary intake had been measured instead, it is doubtful, in view of the difficulty in obtaining reliable measurements of dietary intake over time, whether a dose–response relationship could have been established. The degree of malnutrition in subjects used in such studies might be viewed as a cumulative measure of dietary intake and of effects of factors that influence food utilization over a period of months or even years.

Despite the evidence that malnutrition is associated with behavioral deficits, the extent to which malnutrition, as such, is the cause of behavioral deficits is unclear. The malnourished child is deprived of stimuli required for normal mental and behavioral development because of its inability to participate actively in usual family and societal relationships. Intervention trials have been used to help answer the question of the significance of stimuli deprivation. Cravioto and DeLicardie (1979) in Mexico have shown that nutritional rehabilitation alone is much less effective in bringing about improvement in cognitive and behavioral function than is combined nutritional rehabilitation and sociopsychologic stimulation. Thus, both observational trials and intervention trials have expanded knowledge of relationships between malnutrition and behavior. The extent of genetic influences on behavioral deficits associated with malnutrition, insofar as I am aware, remains to be examined, but Cavalli-Sforza (1984) has presented some suggestive observations that susceptibility to malnutrition may, in some populations, have a genetic component (familial aggregation).

The problems of measuring dietary intake accurately and reliably are discussed by Rhoads. They have been the subject of debate for decades and dealt with again in a recent publication from a committee of the National Research Council (1984). The committee points out that errors arise from two sources: one, from measurements of dietary intake; and the other from errors in food composition values by which measures of food consumption are multiplied to obtain estimates of intakes of individual nutrients.

Rhoads emphasizes the difficulty in validating food intake measurements when there is no standard against which measurements from dietary surveys can be assessed. Guthrie points out in her discussion of Section I that the magnitude of this problem should not be underestimated. Reports of energy and food intakes only slightly above basal energy needs of adult women in the US Health and Nutrition Survey are difficult to reconcile with evidence from health surveys showing a prevalence of overweight of 30%–40% among women. Some of the estimates of reported caloric intakes in other studies in which it has been possible to check them, have been 25%–35% too low. It is clear that the problems of inaccurate reporting of dietary intake, day to day variation in intake, inadequate recall of the consumption of some foods, and inability to check the accuracy of records, make estimates of individual intakes far less reliable than the bold-face figures in tables reporting them would lead us to believe. Variation in the composition of different lots of food, analytic shortcomings, and inadequate measures of bioavailability, compound the problem.

Intake records, recorded daily in diaries, for at least 4 days, and probably 7 to 10 days, increase reliability, but this is frequently difficult to achieve. Above all, training in keeping diet records and motivation to keep them accurately are most critical. Accuracy is improved if a qualified dietitian reviews the diary entries regularly with the subjects keeping the records. For intakes of some nutrients, e.g., protein or sodium, it is possible to check the accuracy of dietary records if 24-hour urine samples are obtained for analysis. Also, the possibility exists of estimating the fat composition of the diet over time if tissue samples can be obtained for analysis. Unfortunately, all of the methods for improving accuracy of intake measurements increase cost and time and are likely to be applied only in studies of small populations.

These problems require continuous attention. The validity of all conclusions from epidemiologic studies of associations between diet and behavior, disease, body weight, health, nutritional status, and other outcomes that may be influenced by intakes of foods or nutrients, depends upon the reliability of the estimates of food intake.

Reliable intake measurements are most critical in studies done in healthy populations of behavioral effects of food components that are not essential, such as carbohydrates. The importance of valid conclusions in these is great because inferences drawn from such studies may be used as the basis for public policy which is difficult to reverse if the conclusions subsequently prove to be wrong. Problems of intake measurement are less critical in studies of effects of malnutrition or specific nutrient deficiencies because, in these, measures of nutritional status provide direct evidence of the adequacy or inadequacy of long-term intakes.

Here we should consider what is implied by the term diet and behavior. As Dews notes, emphasis is currently on associations between intakes of specific nutrients or foods and acute or short-term behavioral responses. We often cite evidence from studies of effects of chronic ingestion of nutritionally inadequate diets to illustrate that severe behavioral responses to diet can occur. This tends to create the expectation that severe effects can be expected from dietary changes within the range usually encountered. In acute, short-term studies, however, behavioral responses that follow diet modifications have been small and subtle. This would be predicted from knowledge of the effectiveness of homeostatic mechanisms, both metabolic and transport mechanisms, that prevent large deviations in the environment of the central nervous system (Harper and Tews 1987). The underlying biological mechanism of behavioral changes observed in chronic studies of effects of inadequate dietary intake and of those observed in acute studies of ingestion of foods or nutrients by healthy individuals are obviously distinctly different. The former can cause permanent defects in brain structures and prolonged impairment of metabolic systems; the latter cause transitory changes in energy supply or neurotransmitter synthesis. It would be well to make a clear distinction between the two by referring to studies of chronic effects of diet, whether dietary inadequacy or excess, as studies of associations between nutritional status and behavior rather than diet and behavior.

Now I should like to turn to the diet–behavior studies in which a quick response is expected. Despite the complexity of behavioral investigations, epidemiologic methods can be applied just as well to studying associations between chronic dietary deficits and behavior (what I have referred to as studies of nutritional status and behavior) as they can to studies of diet and chronic and degenerative diseases. Can they be applied effectively in studies of associations between acute behavioral responses and diet? Rhoads emphasizes the need for experimental approaches in such studies. He elaborates on investigations of diet and hyperactivity, especially those of Harley and associates, in which the controlled experimental approach has been used exclusively to examine associations between ingestion of food additives or sucrose and rapid behavioral responses. In such studies, problems of reliability of dietary intake measures do not pose the formidable task they do in epidemiologic studies because, in short-term studies, dietary intake can be carefully controlled. In fact, in such studies, intake is often observed directly. There are, nonetheless, many other problems that must be surmounted.

It is noted by Vietze that interviews and questionnaires can be used in epidemiologic studies of behavior, just as they are in investigations of food consumption. He and Rodin both discuss problems involved in using them. The logistics of obtaining accurate measures of behavior and dietary intake in a sufficiently large sample size for a valid epidemiologic study, and the difficulty of cross-checking individual answers for reliability, becomes a formidable and prohibitive task. To attempt to determine the incidence of a diet–behavior interaction through interviews and questionnaires would result, I believe, in the collection of a mass of subjective responses, comparable to testimonials. The findings could be assessed for reliability and validity only through carefully controlled experimental studies, which might as well be done in the first place.

The difficulties in achieving an acceptable degree of control of error and bias in classical or traditional epidemiologic studies have been elaborated on by Feinstein (1987) and it is doubtful that his criteria for objectivity, (outlined in the introduction to this section by Simopoulos) could be met in investigations of short-term diet–behavior responses. The double-blind procedure that is essential in controlled experimental studies cannot be applied in epidemiologic investigations. Interviewer bias in asking about events preceding the reported response is difficult to avoid and the bias of the subjects responding cannot be established. Knowledge of a variety of factors such as alcohol intake, drug use and pharmaceutical use, which could influence results, would be difficult to determine and, in any event, assessment of the influence of these agents on the outcome or reported responses would again require the controlled experimental approach. Vietze discusses the problems encountered in matching controls and experimental subjects in order to avoid spurious results. A psychologic evaluation of subjects would be essential in interpreting behavioral responses, and would be needed to determine how the behavioral responses observed were influenced by psychologic traits.

I have interpreted the observations of the three contributors to this section as expressing skepticism about applying epidemiologic methods to the study of the small and subtle, short-term transitory behavioral responses to diet, at least for the present. It would seem, as Rhoads comments, that the controlled experimental approach is the method of choice for such studies. The difficulties involved in using the epidemiologic approach for behavioral testing are emphasized by Vietze's description of the investigations that were done on children exposed to defective formula. As knowledge of the interactions between diet and behavior is expanded, hopefully it will become possible to develop more satisfactory and less elaborate epidemiologic approaches to examine the etiology of behavioral responses to diet.

The need to use the controlled experimental approach in order to advance knowledge of diet and behavior is emphasized by both Rhoads and Vietze. At a meeting on this subject in 1984, Dews (1986) suggested that a field of dietary behavioral pharmacology was needed. He defined this as the study of the behavioral consequences of ingestion of particular dietary components in amounts within the limits of reasonable dietary practices. The problems that will be encountered in drawing a line between this and traditional pharmacology, dietary toxicology, and dietary therapeutics will be more difficult to resolve than he implied. Judgment will be required in deciding, for example, in studies of behavioral responses following tryptophan ingestion, at what level of intake one moves from dietary pharmacology to classical pharmacology to dietary toxicology. This would seem to be a problem with Rhoads' example of the challenge studies that were done with food dyes in large amounts. And might dietary therapeutics not cross all of those lines? Distinctions of this type are important in establishing the legitimacy of extrapolations from results for general applicability. They are not, however, critical in defining guidelines for the methodology.

Sprague (1981), Kreusi and Rapoport (1986), and others have emphasized the need for rigor in studies of diet and behavior, as do the contributors to this section. It has been the failure to evaluate critically the inadequacy of so many of the reported studies on the subject that has led, as Dews (1986) said, ". . . to the pendulum of acceptance of claims for behavioral effects of diet to swing from skepticism to gulibility."

Without going into detail, some of the procedures that need unusually careful consideration when responses are small and transitory should be considered. In selecting the study sample, much more detailed evaluation of subjects, not only for physiologic similarity, but also for similarity of psychologic traits, is required to ensure uniformity of both experimental and control groups. Crossover designs and double-blind techniques must be rigidly maintained to eliminate confounding from placebo effects. Special attention must be given to the selection of the placebo in dietary studies, particularly those in which the substance being tested must be used in large amounts. Even then, increasing one major constituent of the diet will result in sufficient reduction in another to present problems in interpretation. Duration and timing of observations becomes critical in detecting responses. Knowledge of prior nutritional state is essential in assessing responses to diet. Careful standardization of behavioral tests is vital in order to establish the reliability and validity of results. Effects of concomitant, and not readily recognized, variables must be searched out especially carefully when responses are small and transitory. Demonstration of dose-dependency of responses is essential for validation of comparisons between groups. Many of the problems in conducting valid, controlled experimental studies of dietary "pharmacology" are different from those epidemiologic studies, but achieving solutions for them is no less difficult.

On the potential contributions from studies of genetic epidemiology, Ward discusses in considerable detail possible approaches and the problems that must be considered in applying them. A few points should be emphasized.

It is pointed out by Ward that recognition of an underlying metabolic defect, e.g., lactase deficiency, was essential for development of appropriate measures for studies of the "genotype–behavior–environment" interaction in lactose intolerance. The behavioral response acted as a negative feedback system for correction of the genetic problem. Recognition of the association between low resting metabolic rate and obesity has provided a similar tool for epidemiologic studies of the genetic–environment interactions in obesity. In studies of alcohol abuse, as Ward indicates, it was identification of the association of alcoholism with monoaminergically-modulated personality traits that provided the tool for epidemiologic investigations of this problem. For genetic epidemiologic studies of diet and behavior, identification of some similar psychologic trait that is associated with the behavior being investigated would seem to be a necessity. This is no mean task, and again extensive, controlled experimental studies would seem to be essential for providing the tools for the subsequent epidemiologic studies required to unravel underlying genetic components of behavior. If by good fortune, or through arduous effort, such a trait were found to be associated with some well-characterized genetic marker, the rate of progress in elaborating the genetic basis of behavior would be immensely increased. In the meantime, analyzing for familial clustering or aggregation of diet–behavior responses will probably be the most profitable approach for obtaining the necessary clues for more detailed studies.

Garn and associates (1979) have emphasized the utility of such studies in analyzing genetic components that control growth, especially if the studies can be extended to include studies of monozygotic and dizygotic twins, one of whom has been adopted. Ward mentioned the success achieved in using this approach in studies of obesity. An impressive study of twins reared apart has been described by Bouchard and associates (1986) in Minnesota. Their studies of heritability of psychologic traits have so far been limited. Although their findings indicated that not all measures of psychologic traits in twins showed the high correlation necessary to infer strong genetic influence, several certainly did in comparisons with results obtained with cousins.

An additional point emphasized by Ward is the importance of genotype–environment interactions in permitting expression of genes involved in several diseases currently of major interest. I do not think that this can be over-emphasized. Genetic susceptibility to most forms of hyperlipidemia, hypertension, or type II diabetes is not expressed in an environment in which food is sparse, activity is high, salt intake is low, and body weight is low. Ward concludes that expression of the "susceptibility genes" occurs only in a "deleterious" environment. I would suggest that it might be viewed the other way around; that such genes are not expressed in a "deleterious environment," but are expressed when the environment, especially the food supply, is greatly improved and permits high survival rates, achievement of potential body size, and greatly increased life expectancy; hardly a "deleterious" environment except for those who carry the gene. In any event, such observations suggest that genetic influences on behavior might be observed only in specific environments – making it necessary to initiate studies among subjects living in diverse environmental settings.

Finally, why, apart from the desire to expand basic knowledge of psychobiological phenomena, is it important to undertake behavioral and genetic epidemiologic studies of diet and behavior when, as Kreusi and Rapoport (1986) have concluded, the behavioral responses to diet in short-term studies are small, transitory, and not of clinical significance? Feinstein (1987), after presenting a critical evaluation of epidemiologic methods, and pointing out that it is still an underdeveloped science, emphasized that, although much remains to be done to improve the methodology, it is the only feasible method for studying the etiology of human diseases. In discussing dietary behavioral pharmacology, Dews (1986) emphasized that the myths about diet and behavior pose a threat to science. Unless the false claims and unwarranted generalizations are examined critically and their lack of validity is exposed, acceptance of such claims can lead to public policy decisions being based on pseudoscience. And last but not least, even if diet modifications within the "normal" range are not considered to be of potential clinical significance, understanding of such observations still may lead the way to developing dietary therapeutic approaches, possibly with pharmacologic levels of nutrients, for treatment of certain behavioral problems or diseases.

References

Agarwal DK, Upadhyay SK, Tripathi AM, Agarwal KN (1987) Nutritional status, physical work capacity and mental function in school children. Nutrition Foundation of India, New Delhi, scientific report no. 6, pp 1–86

Bouchard Jr TJ, Lykken DT, Segal NL, Wilcox KJ (1986) Development in twins reared apart: a test of the chronogenetic hypothesis. In: Demirjian A (ed) Human growth: a multidisciplinary review. Taylor and Francis, London, pp 299–310

Cavalli-Sforza L (1984) Future needs and developments: a geneticist's point of view. In: Valazquez A, Bourges H (eds) Genetic factors in nutrition. Academic Press, New York, pp 423–434

Cravioto J, DeLicardie ER (1979) Stimulation and development of malnourished infants. Lancet ii:899

Dews PB (1986) Dietary pharmacology. Nutr Rev 44[Suppl]:246–250

Feinstein AR (1987) Scientific standards and epidemiologic methods. Am J Clin Nutr 45:1080–1088

Garn SM (1979) Genetic and nutritional interactions. In: Alfin-Slater RB, Kritchevsky D (eds) Human nutrition, vol 2, Nutrition and growth. Plenum Press, New York, pp 31–46

Harper AE, Tews JK (1987) Nutritional and metabolic control of brain amino acid concentrations. In: Heuther G (ed) Amino acid availability and brain function in health and disease. Springer, Berlin Heidelberg New York, pp 3–12

Kreusi MJP, Rapoport JL (1986) Diet and human behavior: how much do they affect each other? Ann Rev Nutr 6:113–130

National Research Council (1984) Nutrition adequacy. National Academy Press, Washington DC

Sprague RL (1981) Measurement and methodology of behavioral studies: the other half of the nutrition and behavior equation. In: Miller SA (ed) Nutrition and behavior. The Franklin Institute Press, Philadelphia, pp 269–275

Section IV

Integration of Research Methods in Diet and Behavior

Introduction

G. D. Miller

The study of diet and behavior is a relatively young field and methodologic concerns will increase as study in this area continues. For the field to progress it is essential that scientists gain an understanding of the appropriate use of methods from other disciplines and effectively integrate these methods into their area of study. Although many methodologic issues are considered throughout this book, an ongoing dialog between scientists will be needed in order to resolve new methodologic issues as they develop. The incorporation of methods from many other disciplines, such as pharmacology, toxicology, neurophysiology, anthropology, etc., will occur. This multidisciplinary approach will allow us to answer the most difficult questions. However, with many disciplines involved in answering questions, misunderstandings and lapses in communication may occur. An intense effort by all scientists involved in diet and behavior research to communicate new results and methods, as they develop, will help alleviate such confusion and misunderstanding. This will allow effective utilization and integration of methods from many different disciplines. In a volume that brings together scientists interested in the interaction between diet and behavior it is possible to consider the strengths, weaknesses, and utility of the various methods used to study this interaction, as well the appropriate interpretation of the results of such studies. Such a book would not be complete without a section dedicated to the integration of methods used by behavioral and nutrition scientists. An effective integration of nutrition and behavioral methods and the appropriate interpretation of results will occur if there is clear communication and continuing dialog between the two disciplines. In this section the contributors of the first two chapters integrate and place into perspective methods used for diet and behavior research. Through these different perspectives a better understanding of how to integrate and utilize methods developed by the different disciplines can be gained. In the concluding discussion M. R. C. Greenwood examines future needs and challenges which, if resolved, will allow the study of diet and behavior to grow through a greater interaction of different disciplines.

In her paper C. Greenwood examines the physiologic psychologist's approach for studying the interactions between diet and behavior. In this approach, the experimental design has an emphasis on understanding the mechanism involved in how changes in diet alter behavior. Establishing the physiologic mechanisms allows for more appropriate selection of behaviors to test and establish if the influence of diet is functionally relevant and under what condition (e.g., level of nutrient, length of exposure). Conversely Greenwood points out that utilizing the observations of behavioral scientists can be important in developing new hypotheses on potential mechanisms involved in diet–behavior relationships. To gain a better understanding of the influence of diet on behavior, sample size calculation, quantification of diet, use of food composition tables, range of nutrient intake, and duration of feeding are considered in the framework of integrating different disciplines.

In their paper Rolls and Hetherington discuss how studies of diet and behavior can benefit from the use of methods developed by behavioral scientists. The authors clearly demonstrate how method selection may change as the experimental question varies. Utilizing various examples they illustrate a range of experimental approaches that can be undertaken to examine how diet influences behavior. The results of studies examining food intake and selection, as influenced by various sweeteners, illustrate laboratory methods employed to understand experimental and hedonic aspects of eating behavior. The studies of sensory-specific satiety conducted by Rolls are excellent examples of the variety of approaches that can be undertaken to study diet and behavior relationships.

The concluding chapter by M. R. C. Greenwood discusses the need to develop new methods and standards to meet some of the challenges and problems of multidisciplinary interaction raised in this volume. She suggests that establishing more effective communication networks may be helpful in meeting new research challenges. Development of new animal models may help in understanding genetic expression and its influence on diet–behavior interactions. She also emphasizes that no one person can be cognizant of all methods in other disciplines and reminds us of the need to develop integrative thinking in the next generation of scientists so that future research will incorporate information and methods from many different disciplines.

Chapter 10

Methodologic Considerations for Diet and Behavior Studies: A Nutritionist's Perspective

C. E. Greenwood

Introduction

The relationship between diet and behavior provides a basis of scientific investigation for a number of disciplines. The interdisciplinary nature of the research field allows for exciting, innovative and productive approaches to address questions of a multifaceted nature. However, as is true with all interdisciplinary fields of study, it is imperative that exacting attention be given to experimental design details such that the needs and rigors of all disciplines are met.

Initial consideration must be given to the level of organization at which both diet and behavior are being assessed. Diet can be viewed as either the combination of nutrients or the combination of foods being consumed by an individual or population. Behavior, on the other hand, can be viewed as a single outcome measure, for example, total food consumption, or as a sum of all components of that behavior, such as rate and intensity of eating or patterning of food and/or nutrient selection within a meal or throughout a longer time-frame. Measures of diet and behavior will depend upon the experimental question being asked. In general, studies of the metabolic and/or physiologic impact of food consumption are best done at the nutrient level. In contrast, studies examining the impact of psychosocial factors on food selection are best done at the food level. Similarly, studies aimed at elucidating the metabolic control of behaviors, such as food selection, will concentrate only on those aspects of the behavior that are under physiologic control. However, a full understanding of human behavior can only be gleaned by elaborating all aspects of behavior including both the physiologic and non-physiologic components. The choice of variables used and measurements made must, in all cases, be hypothesis driven. It is essential that experiments are not designed in a void and that consideration be given to other facets of diet and behavior methodology. In many cases, this will involve recognition of and control for variables not under examination.

The intent of this and the following chapter is to provide guidance in experimental design considerations for studies into diet and behavior. The following chapter addresses issues related to behavioral assessment; this chapter will address issues related to the dietary aspects of these studies. An overview of two common disciplinary approaches to studies of diet and behavior will first be provided, followed by specific design features to be considered, including sample size calculations, quantification of the diet, use of food composition tables, range of nutrient intakes, and duration of feeding.

Experimental Approaches

For the purpose of illustration, experimental aproaches have been divided into physiologic psychology and behavioral sciences. This by no means represents a definitive separation, but does allow for a comparison which highlights some of the research needs within each area.

Physiologic Psychology

In general, the physiologic psychologist is interested in understanding the relationships among diet, brain metabolism and behavior. Probably the two most important distinctions to be made in this approach are first that diet is addressed at the level of nutrients rather than of foods. So the form in which the diet is presented is relatively unimportant unless either the metabolic impact of cephalic phase response or altered rates of absorption is one of the criteria under consideration. The second distinction is that by and large only the physiologic component of behavior regulation is under study. The underlying assumption in this approach is that there is a continuum between nutrient intake, altered brain metabolism and changes in behavior. That is, if brain metabolism is altered via dietary means, then behavior will be changed in a predictable manner. A component of this assumption is that a dose–response relationship exists not only between diet (nutrient intake) and brain metabolism, but also between brain metabolism and behavior.

To date, three different mechanisms have been identified whereby diet can influence brain metabolism (Fig.10.1). These mechanisms have been reviewed in detail elsewhere (Leprohon-Greenwood and Anderson 1986; Greenwood and Craig 1987) and will only be briefly summarized here. First, it is known that the synthesis of at least five neurotransmitters is under precursor control (i.e., the rate-limiting enzymes involved in their synthesis are not saturated under either quiescent and/or activated conditions). Included amongst these neurotransmitters are serotonin, the catecholamines (dopamine and norepinephrine), histamine, glycine and acetylcholine, which are synthesized from tryptophan, tryosine, histidine, threonine and choline respectively. Dietary manipulations which increase neuronal concentrations of these precursors may lead to increased synthesis and release of the neurotransmitter (reviewed in Wurtman et al. 1981). This relationship must be qualified, in that the firing rate of the neuron may play an important role. This has been clearly demonstrated for the catecholamine neurotransmitters. Precursor dependency of the catecholamines is usually only observed when the neuron is actively firing. This may be related to altered affinity of the rate-limiting enzyme, tyrosine hydroxylase, for its substrate, tyrosine

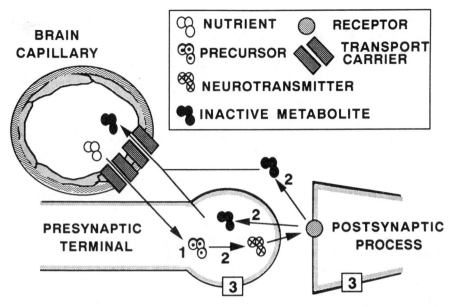

Fig. 10.1. Proposed mechanisms by which diet influences brain metabolism. 1: Increased availability of precursors for the neurotransmitters serotonin, the catecholamines (dopamine and norepinephrine), histamine, glycine and acetylcholine will increase their syntheses since the rate-limiting synthetic enzymes are not saturated under normal physiologic conditions. 2: Since vitamins and minerals serve as cofactors for synthetic and degradative enzymes involved in neurotransmitter metabolism, deficiency will result in altered neurotransmitter metabolism. 3: Dietary supply of fatty acids can alter membrane composition which in turn may influence membrane-associated events.

(reviewed in Fernstrom 1983; Sved 1983), and/or intraneuronal depletion of tyrosine in catecholaminergic neurons when synthetic demand is high (Milner and Wurtman 1985).

A second mechanism is alterations in vitamin and mineral availability (see Sandstead 1986; Dreyfus 1988 for reviews). This relationship is observed only when the micronutrients are provided below physiologic requirement and relates to the fact that micronutrients serve as important cofactors for the synthetic and degradative enzymes involved in neurotransmitter metabolism. In addition, metaloproteins (which may not be directly involved in neurotransmitter metabolism) will also be decreased by mineral deficiencies. For example, in examining the impact of iron deficiency, little change in the metabolism of the monoaminergic neurotransmitters is observed, despite the fact that both the synthetic and degradative enzymes are iron-requiring (Youdim et al. 1980). However, decreased quantities of certain other proteins are observed (Youdim et al. 1986) including the D_2-dopamine receptor (Youdim et al. 1983). The complete profile of proteins influenced by iron deficiency has not been identified. Nor is it known at present whether these proteins are in themselves iron-containing or if they require iron-dependent enzymes for their synthesis. However, alterations in their synthesis may play a fundamental role in the behavioral sequelae associated with iron deficiency.

Finally, it has been established that the composition of brain membranes is sensitive to the supply of fatty acids. This was first observed in models of essential fatty acid deficiency where it is not surprising that reduced quantities of essential fatty acids and their metabolites would be observed in membrane phospholipids from deficient animals. However, it now appears that brain membrane phospholipid fatty acid profile is sensitive to the dietary supply of fatty acids even in the absence of essential fatty acid deficiency, in a manner analogous to that reported for peripheral tissues (Foot et al. 1982; McGee and Greenwood, 1989). The magnitude of change, however, may not be as great as that observed in the periphery (Matheson et al. 1981). Alterations in membrane composition may in turn impact on brain function mediated either indirectly via alterations in membrane-bound protein function or directly via alterations in receptor mediated phospholipid turnover (for reviews see Gould and Ginsberg 1984; Shinitzky 1984; Stubbs and Smith 1984).

Given an understanding of these relationships, researchers in this area will turn to behavioral approaches for one of two purposes. First, if nutrient intake impacts on central nervous system (CNS) metabolism, it is essential to ask whether these metabolic changes are functionally relevant. To address this question, behaviors will be selected based on the known neurotransmitter pathways (or other metabolic events) involved in their expression, as a method of addressing physiologic relevance. The interest here is not always in the behavior *per se*, but rather as using the behavioral measure as a means of validating the relevance of metabolic alterations. For example, doses of tryptophan within the range of daily food ingestion as well as carbohydrate meals have been shown to increase brain serotonin levels in experimental animals. This has been indirectly confirmed in humans by measuring increased concentrations of the serotonin metabolite, 5-hydroxyindole acetic acid, in ceresbrospinal fluid following tryptophan administration. To determine the functional relevance of increased brain serotonin concentrations, a number of serotonin mediated behaviors, including food intake regulation, pain sensitivity and aggression have been examined in both experimental animals and humans. In addition, the efficacy of pharmacologic levels of tryptophan administration in the treatment of human disorders such as depression and insomnia has been explored (see Fernstrom 1983; Sved 1983; Young 1986 for reviews). Results from these studies provide evidence that nutrient-induced alterations in serotonin metabolism can indeed impact on behaviors mediated in part by serotonin.

Conversely, other studies in this area are aimed at trying to elucidate the neurochemical basis of certain behaviors. There is a large body of research examining the role of various neurotransmitters in food intake regulation and selection. Experimental approaches are generally aimed at altering the transmission of specific neurotransmitters as a means of testing whether these neurotransmitters are involved. These approaches may use either nutrients themselves, or drugs which are known to alter the transmission of neurotransmitters (see Wurtman and Wurtman 1984: Anderson 1988 for reviews). However, the common aim is to understand the neurochemical basis of the behavior, with the long-term goal of being able to influence behavior in a predictable and desirable fashion.

Behavioral Sciences

Probably the biggest distinction that can be made here is once again at the level of discrimination of the dependent and independent variables. In most behavioral science studies, diet is thought of in terms of foods, so that not only the nutrient composition of the food, but also other attributes including taste, texture and appearance are taken into consideration. Furthermore, interest in behavior is not confined to its physiologic control, but other non-physiologic factors are taken into consideration (see Chapter 1).

While methodologic concerns for the behavioral sciences are addressed in the following chapter (Chapter 11), there are a number of important issues that the physiologic psychologist can learn from these studies. First and foremost is the fact that the non-physiologic components of behavior must be addressed, or controlled for, in experimental designs. An appreciation of the non-physiologic aspects may help to explain why some data are internally consistent within laboratories but are not reproducible from one laboratory to another. For example, in preparing diets for experimental animals, diets of identical nutrient content may be presented to the animal in either pellet, powdered or gel form. The impact of the texture (and perhaps taste) of these diets in animal experimentation is often overlooked. Similarly, in human experimentation, providing a nutrient in pill form, as an artificial preparation (e.g. slurry) or as a traditional food may impact on the outcome of the experiment (see Chapter 4). Providing a nutrient as a pill will bypass the normal cephalic phase response and the physiologic sequelae associated with it. Therefore, it may be appropriate to use tablet/pill forms of nutrients if the question under study specifically relates to non-cephalic phase effects of that nutrient; however, results from studies in which a nutrient has been incorporated into food may not be directly comparable with results from studies in which the nutrients have been provided in an artificial format.

Methodologic Considerations

Sample Size Calculations

The importance of sample size is an issue which has been frequently overlooked in both experimental design and data interpretation. When inappropriate sample sizes are used, the interpretation of experimental data can be meaningless (Anderson and Hrboticky 1986). If sample size is too small, even large treatment effects may fail to be significantly different; conversely, with large sample sizes extremely small treatment effects can be statistically significant. That is, samples that are too small can prove nothing, whereas samples that are too large can prove anything. In behavioral experiments, because of the potentially extensive sources of extraneous errors in both the independent and dependent variables, the risk of obtaining false negative results (type II error) is large. This problem can be minimized, however, with the use of adequate sample sizes.

To predetermine the correct sample size, several criteria of the design have to be known or predicted. These include the population variability in the dependent variable (usually the behavioral outcome), the desired or predicted size of the expected outcome, and the desired probability of false-positive and false-negative results. The latter two criteria are conventionally set at $P > 0.05$ and $P > 0.20$, respectively. Methods for calculating sample size can be found in a variety of sources (e.g. Rosner 1986; Cohen 1988).

Impact of Small Sample Sizes

When sample sizes are inappropriately small, data will lack statistical significance. Care in determining adequate sample sizes must be taken in the following two circumstances. First, in a model where the predicted link between behavior and diet is weak (but still considered physiologically significant), small changes in the behavioral outcome should be expected with dietary manipulations. Under these circumstances, large sample sizes are necessary to demonstrate statistical significance. Second, large sample sizes are necessary when the variance of the dependent variable is large. Studies should not be performed unless a sufficient sample size to obtain statistically relevant results can be run since studies in which sample size is too small will produce meaningless data.

Impact of Large Sample Sizes

Interpretative difficulties associated with sample sizes which are too large are generally observed with correlation studies involving large data bases. These studies often run into the danger of making false-positive (type I error) conclusions. That is, in studies where a large number of correlations are performed on the same data base, the likelihood of obtaining statistically significant correlations increases with the number of subjects and with the number of associations run. Under these conditions, small correlations become statistically significant, and it is then up to the investigators to decide on the physiologic relevance of such results. Furthermore, chance alone will dictate that up to 5% of the associations tested will be declared statistically significant. An example of this can be seen by examining the results of a study of 260 free-living elderly people in which the intake and blood levels of several nutrients were correlated with the results of several cognitive tests to determine the impact of nutritional status of the elderly on cognitive performance (Goodwin et al. 1983). Significant correlations were found for two out of the 24 associations run. Although statistically significant, the correlations were small, accounting for only 2%–3% of the variance of the cognitive tests. The relevance of such findings should therefore be open to question.

In cases where a small number of correlations are run, Bonferroni's adjustment (Cupples et al. 1984) is one way to overcome the likelihood of making a false-positive error. This adjustment divides the statistic obtained by the number of correlations run. For example, if $P>0.05$ was the predetermined α-error and 5 correlations were run, the level of significance would be taken at 0.01 (to be equivalent to $P>0.05$). The problem with this adjustment is that as the number of correlations run increases, the acceptable probability level becomes unrealistically low. For example, if 100 correlations were run, a level of significance of 0.05/100 or $P>0.0005$ would be taken (to be equivalent to $P>0.05$).

Quantification of Diet

Quantification of diet has been discussed earlier in this volume (see Castonguay and Stern, Appendix to Section II) and there is no need to reiterate the detail here. One area, however, which deserves further attention is the impact of nutrient dilution. This has been problematic in studies where nutrients (especially fat) have been added to standard laboratory chow in an attempt to study the nutrient's effect. In some instances, fat was added to chow prior to feeding the animals. This effectively reduces the nutrient density of the diet, such that animals may become deficient in other nutrients. It is essential to maintain the density of nutrients relative to the caloric content of the diet. Furthermore, he need for certain nutrients, such as vitamin E, is better expressed relative to the intake of polyunsaturated fatty acids, and hence its requirement increases as the proportion of fat in the diet increases.

Another instance where diet manipulation has been inappropriate is the addition of high levels of non-nutritive fillers (such as cellulose) to diets to examine the animal's ability to increase food intake to maintain energy balance. While this approach can be extremely useful, it is important to recognize that with extremely high fiber levels, the food becomes bulk limiting such that the animals are physically incapable of a further increase in their consumption and intake results become meaningless.

Use of Food Composition Tables

Many different computerized data bases or tables for food composition are available. These data bases have an enormous utility in nutritional research; however, one must be aware of their limitations and restrict their use to appropriate circumstances (Rand et al. 1987).

The major limitation of these data bases is that they provide average nutrient contents for the foods listed. There is considerable variability from one food to another depending upon such factors as agricultural and manufacturing conditions. The mineral content of any agricultural product will vary depending upon the geographical area (and soil conditions) where the crop was grown. Furthermore, the source of certain major ingredients, such as fats and oils, will vary in most foodstuffs due to alterations in marketplace prices and availability. If one is trying to assess the average intake of a nutrient in a population, this variance is probably not of great concern. However, if one is trying to assess the intake of a nutrient for an individual, a considerable degree of error may be introduced (Beaton 1987).

This may be especially problematic in studies where nutrient selection from single meals is being examined. There are numerous studies where the acute effect of a nutrient or drug on meal selection by an individual is being studied (e.g., Hrboticky et al. 1985; Ryan-Harshman et al. 1987; Young et al. 1988). These designs often provide experimental subjects with a cafeteria-style meal where a limited number of foods is provided in excess. Selection from the different foods is measured and nutrient intake is predicted based on food composition tables. In this situation, the error associated with nutrient estimates may be great enough to mask the effect under study. When limited numbers of foods are involved, it is preferable to analyze the nutrients of interest and use these values for further calculations.

Range of Nutrient Intakes

In simple terms, the range of nutrients used in an experimental paradigm can be categorized into deficient (i.e., below current dietary recommendations), normal (i.e., within the range of human consumption patterns but above requirement) and supraphysiologic (i.e., above the level found in the human diet). Obviously in the study of the impact of diet on behavior in humans, the most important question is whether normal variations in food intake result in behavioral abnormalities. Unquestionably, providing a diet inadequate in nutrients for a long enough period of time to deplete body stores and tissue levels of those nutrients will impact on metabolism and in most cases CNS metabolism. It is therefore not unreasonable to think that brain function or behavior will be adversely affected by these nutrient deficiencies. Conversely, use of excessively high doses of nutrients which exceed levels obtainable from the food supply tell us little about whether diet influences human behavior, since an individual would not be exposed to this level of intake.

This does not say, however, that designs using deficiency or supraphysiologic doses are always inappropriate. Table 10.1 outlines examples where the use of these paradigms may be used.

Table 10.1. Range of nutrient intakes and their value in studies of diet and behavior

Range	Definition	Value
Deficiency	Below dietary recommendations	Generally not applicable unless range of deficiency examined is observed in the human diet Impact of supplementation useful
Normal	Within range of human consumption patterns, but above requirement	Most useful range to assess relationship between diet and behavior
Supraphysiologic	Above level found in human diet	Not applicable to human situation Advantageous for initial pilot studies May have clinical/therapeutic value

Nutrient Deficiencies

Nutrient deficiency paradigms are most useful when the degree of deficiency imposed is one which is observed in the human diet. To illustrate this, results of some intriguing and potentially critical studies regarding the impact of iron deficiency on human cognitive performance can be examined (Oski et al. 1983; Pollitt et al. 1986). These studies demonstrated that cognitive performance of infants and children with low iron stores, but not necessarily with iron deficiency anemia, showed impaired cognitive performance relative to age-matched iron-replete controls. That iron depletion was at least one of the factors involved in the poor cognitive performance was demonstrated by an improvement in cognitive performance with iron supplementation. What makes these studies so important are two issues. First, the degree of iron depletion is one commonly observed in the human population both in developed and developing countries. Second, there appeared to be an association between cognitive performance and iron status that was observed in infants and children not showing overt signs of anemia; however, in routine medical practice, unless there are other underlying reasons, only hemoglobin measures are taken. Thus these data suggest that nutrient-induced impairment may exist which would not ncessarily be picked up by normal screening practices.

Supraphysiologic Doses

The value of using supraphysiologic doses comes from one of two reasons. First, in designing a pilot study for examining the effect of a nutrient on behavior in which there is no pre-existing information, the supraphysiologic dose provides a rapid method of assessment. That is, if the behavior is not sensitive to high doses, it is unlikely that it will be sensitive to lower doses of that nutrient. An example of the appropriate use of supraphysiologic doses can be seen in the following study designed to examine the behavioral effects of aspartame (in tablet form) in healthy adult male volunteers. Doses of aspartame equivalent to 70 to 100 cans of soda were not found to affect measures of mood, arousal, or total food intake and macronutrient selection (Ryan-Harshman et al. 1987). Therefore, it would seem unreasonable to predict that lower doses of aspartame would impact on any of the above measured parameters. (It is important to recognize that this study was designed to examine the effect of elevating plasma and brain levels of the amino acid phenylalanine on behavior. Hence the cephalic phase response to aspartame was bypassed by providing the aspartame in tablets. These data therefore cannot be used to address the effect of stimulation of taste receptors, or other components of cephalic phase response, on the parameters measured.) The supraphysiologic dose provides a rapid pilot examination. However, if a behavioral effect is observed, it is essential that the experiments be repeated using normal doses of the nutrient before the results can take on any meaningful physiologic relevance.

Second, the use of supraphysiologic doses of nutrients may play a role in the treatment of disease. Following the observation that tryptophan administration increases brain serotonin levels, the efficacy of high doses of tryptophan in the treatment of a number of disorders including depression, insomnia and obesity, has or is being examined (reviewed in Wurtman and Wurtman 1984; Young 1986). This scenario represents the pharmacologic application of nutrients and while it may play an important medical role, the results should not necessarily be extended to the normal human diet.

Normal Doses

In designing studies examining normal variations in intake, several criteria can be used (see Schwab and Conners 1986). In general, for studies relating to nutrients, the doses used should span the range of that normally observed in the diet. This range may be based either on average consumption patterns or on ranges observed within single foods or meals. Selection of criteria must be based on the experimental question under examination. Studies of a chronic nature (see below) are most often based on average consumption patterns, while acute studies (i.e., effects of a single bolus of a nutrient or of a single meal) can be based on either average consumption or on single foods or meals.

Studies designed to examine the impact of non-nutritive components of the diet (for example food additives) once again should be based upon anticipated human exposure level. The common criterion used for these studies is either the acceptable daily intake (ADI) or a predicted abusive dose (calculated assuming an excessive intake of foods containing the additive). In all situations it is preferable to administer the nutrient or non-nutritive dietary component on a weight/kg body weight basis to control for different body weights of the experimental subjects.

Acute vs. Chronic Studies

The choice of whether to use a chronic or acute feeding paradigm must be hypothesis driven and based upon the known or expected physiologic mechanism(s) involved and the time-course of the physiologic change. Studies designed to elucidate the neurochemical pathways involved in meal to meal regulation of food intake, can best be examined using acute paradigms where the testing period only extends over one or several meals. In contrast, studies designed to examine the impact of vitamin and mineral deficiencies or dietary fat-induced alterations in membrane composition should use chronic paradigms which span the expected time-course for metabolic changes to occur as a result of the diet manipulations. Examples of the use of these two paradigms are outlined in Table 10.2.

Table 10.2. Study duration and proposed physiologic mechanisms

Studies	Duration	Nutrients studied	Physiologic mechanisms
Acute	Minutes to hours	Amino acids, choline	Altered neurotransmitter metabolism due to changes in precursor availability
Chronic	Days to months	Vitamins	Altered neurotransmitter metabolism due to decreased enzyme acitivity
		Minerals	Altered neurotransmitter metabolism due to decreased enzyme activity Decreased metaloprotein synthesis
		Fatty acids	Altered membrane composition

Acute Studies

Studies using acute paradigms most commonly examine the behavioral impact of altered neurotransmitter precursor availability and hence neurotransmitter synthesis. Alterations in brain amino acids and neurotransmitter levels are observed within 15 – 30 minutes after administering amino acids or feeding diets high in either carbohydrate or protein to experimental animals (Fernstrom et al. 1973; Li and Anderson 1982, 1984; Ablett et al. 1984). These neurochemical changes are evident up to several hours after the experimental treatment depending upon dosages used. Thus behavioral testing should be conducted within the time-frame of the expected metabolic change. In studies designed to determine the neurochemical basis of food intake regulation, a number of investigators have used acute paradigms (single meal or meal-to-meal) to elucidate neurochemical signals involved in the short-term regulation of both quantitative (total meal consumption) and qualitative (proportions of macro-nutrients selected) aspects of food intake regulation. Results from these studies in experimental animals have demonstrated that meals high in carbohydrate result in the subsequent selection of meals high in protein and vice versa (Li and Anderson 1982). Furthermore, brain levels of serotonin are inversely related to the proportion of carbohydrate contained within a meal. By combining the results of these studies with similar observations in both experimental animals and humans following the administration of either tryptophan or drugs known to enhance serotonin neurotransmission, a hypothesis that serotonin is at least one of the factors involved in this regulation has evolved (Anderson 1988; Wurtman and Wurtman 1984).

Obviously, data derived from chronic feeding of diets cannot be used to interpret short-term regulation of food intake. An inappropriate use of chronic feeding for the assessment of meal to meal regulation of food intake can be seen in studies where diets varying in protein and carbohydrate concentrations were fed chronically to rats (11 – 14 days) and brain amino acid and neurotransmitter levels monitored (Fernstrom et al. 1985; Peters and Harper 1985). Little information can be gleaned regarding meal-to-meal regulation of food intake because sacrifice times did not coincide with either meal initiation or termination. That is, in order to assess the involvement of neurotransmitters in behavioral regulation it is essential that biochemical measurements coincide with relevant aspects of that behavior.

Chronic Studies

The use of chronic paradigms is more aptly applied to studies of either micronutrient deficiencies or alterations in the composition of dietary fat. In both instances, longer-term feeding of the experimental diet is necessary for the anticipated neurochemical change(s) to occur. Results from acute studies would be difficult to interpret unless new metabolic mechanism(s) of action were being proposed.

To illustrate this point, an example from studies examining the behavioral effects of altering the composition of dietary fat will be examined. In these studies, it was hypothesized that dietary fat would alter brain function and behavior mediated via alterations in membrane composition. Animals were exposed to different dietary fats (soybean oil vs. lard or beef tallow) for 2 – 3 weeks prior to and throughout the behavioral testing in order to provide sufficient time for membrane compositional changes to commence. Following this period, it was observed that protein and carbohydrate intake regulation was altered such that rats fed the soybean oil diet selected less protein and more carbohydrate than rats fed either lard or beef tallow diets (Crane and Greenwood 1987; McGee and Greenwood 1989). In a subsequent experiment, in which dietary selection commenced at the same time as the dietary fat manipulation, the alterations in protein and carbohydrate selection did not reach statistical significance until 14 – 18 days' exposure to the fat sources (McGee and Greenwood, 1989). Thus these results are consistent with the hypothesis that metabolic changes associated with the dietary manipulation must occur prior to the expression of behavioral change.

Blending of the Disciplines: Collaboration not Duplication

Obviously in multidisciplinary fields it will only be the exceptional few who master the tools of all trades. Therefore the challenge is to integrate research findings from other disciplines into our own framework. This can be accomplished either by developing new hypotheses based on existing information or preferably by active collaboration with researchers versed in complementary methodology. Two such examples are outlined below.

First, in examining the work of Drewnowski and co-workers (1985), it is apparent that hedonic taste responses of obese, formerly obese and normal weight individuals differ such that obese individuals show a preference for solutions higher in fat and lower in sucrose in comparison to normal weight individuals, while the formerly obese prefer higher fat (similar to the obese) and higher sucrose (similar to the normal weight) solutions. One simplistic hypothesis that can be derived from these data is that sucrose preference is related to the degree of adiposity while fat preference has a more genetically controlled base since it does not change with normalization of body weight. Provided that this hypothesis is sound, one logical experimental avenue may be to determine if the metabolic (or molecular) basis of signals controlling fat preference is involved in the etiology of obesity. An important extension of this would then be to determine whether metabolic or molecular measures can be developed as a method of screening infants or children at risk for obesity so that intervention can be applied prior to the expression of obesity. Obviously these questions will not be answered in the near future.

However, a more immediate and progressive avenue may be in the development of specialty food products which will aid in dietary compliance of obese individuals attempting to lose weight. That is, low calorie foods should be developed to comply with taste preferences of obese, rather than normal weight, individuals. With the introduction of fat substitutes into the market place, it will soon be possible to provide high "fat"/low sucrose foods which are highly palatable to the obese individual yet low in caloric density. A multitude of other research avenues can also be evolved from the initial observations of Drewnowski and colleagues (1985); however, the message here is that the physiologic psychologist should not expend energy reproducing the original findings (those equipped with sophisticated behavioral sciences methodology are better trained to do this), rather, effort should be placed on the evolution of new hypotheses founded on behavioral sciences observations.

Secondly, the greatest degree of sophistication in research approaches is usually achieved when individuals with complementary training and methodologic expertise actively collaborate. Generally, investigators initially trained in nutrition and metabolism rely on behavioral tasks commonly used by behavioral scientists. Unfortunately, many of these tasks were designed to assess the impact of relatively large alterations in brain function, such as electrolytic lesions to specific brain regions or pharmacologic alteration of transmission. In contrast, the degree of metabolic change associated with dietary manipulations is not as great as that produced by lesioning or pharmacologic techniques; hence the behavioral tasks may not be sensitive enough to assess the physiologic relevance of diet manipulation. However, as discussed by Thompson et al. (Chapter 6), those with expertise in behavioral methodology are best able to design testing procedures with the appropriate degree of sensitivity to assess relatively small changes in metabolism. An illustration of the advantages of collaboration comes from our investigation of the impact of altered dietary fat composition on brain function. Early experiments demonstrated that dietary fat (20% w/w soybean oil vs. lard diets) influenced cognitive performance in rats assessed using the Morris place navigation tank (Coscina et al. 1986), with rats fed the soybean oil diet showing superior performance in comparison with those fed the lard diet.

This observation was of limited value in isolation. In addition, the sensitivity of the task to the dietary manipulation was not great enough to demonstrate reliable effects in subsequent experiments. Through the use of a variety of measures of learning and memory, including radial arm and Hebb-Williams mazes and a variable-interval delayed alternation task, we were able to demonstrate that learning and memory are indeed influenced by dietary fat. Results from this study suggested that the impairment was not confined to a single brain region (such as the hippocampus). Rather, in tasks requiring the integration of information from a number of brain regions, such as the Hebb-Williams mazes and the variable-interval delayed alternation task, greater impairment was observed in lard-fed animals suggesting that a number of brain regions were involved (Greenwood and Winocur, in press). The simple and obvious message is that through the integration of different disciplines, experimental protocols can be designed which capitalize on expertise in both dietary and behavioral methodology.

Conclusions

The dichotomy that will continue to persist in studies of diet and behavior is that individuals consume foods while the body sees nutrients. In general, studies examining the physiologic relationships among diet, brain metabolism and behavior are best done at the nutrient level and must be hypothesis driven based on known or proposed metabolic mechanism(s). While the behavioral measures in these studies are often confined to physiologic aspects of behavior, care must be taken to control for the non-physiologic components of the behavior under study. In contrast, a complete understanding of human behavior, including food selection, can only be gleaned by elucidating the interrelationships among all components of the behavior in question.

If the long-term goal is to alter human behavior, based on metabolic and physiologic information, this can only be effectively accomplished by providing advice and alternative avenues which are consistent with all aspects and components of the behavior involved.

References

Ablett RF, MacMillan M, Sole MJ, Toal CB, Anderson GH (1984) Free tyrosine levels of rat brain and tissues with sympathetic innervation following administration of L-tyrosine in the presence and absence of large neutral amino acids. J Nutr 114:835–839

Anderson GH (1988) Metabolic regulation of food intake. In: Shils ME, Young VR (eds) Modern nutrition in health and disease, 3rd edn. Lea and Febiger, Philadelphia, pp 557–569

Anderson GH, Hrboticky N (1986) Approaches to assessing the dietary component of the diet-behavior connection. Nutr Rev 44[Suppl]:42–50

Beaton GH (1987) Consideration of food composition variability: what is the variance of the estimate of one-day intakes? Implications for setting priorities. In: Rand WH, Windhan CT, Wyse BW, Young VR (eds) Food composition data: a user's perspective. The United Nations University, The Bath Press, Avon, UK, pp 194–205

Cohen J (1988) Statistical power analysis for the behavioral sciences, 2nd edn. Lawrence Erlbaum, Hillsdale, New Jersey

Coscina DV, Yehuda S, Dixon LM, Kish SJ, Leprohon–Greenwood CE (1986) Learning is improved by a soybean oil diet in rats. Life Sci 38:1789–1794

Crane SB, Greenwood CE (1987) Dietary fat source influences neuronal mitochondrial monoamine oxidase activity and macronutrient selection in rats. Pharmacol Biochem Behav 27:1–6

Cupples LA, Heirn T, Schatzkin A, Colton A (1984) Multiple testing of hypotheses in comparing two groups. Ann Intern Med 100:122–129

Drewnowski A, Brunzell JD, Sande K, Iverius PH, Greenwood MRC (1985) Sweet tooth reconsidered: taste responsiveness in human obesity. Physiol Behav 35:617–622

Dreyfus PM (1988) Vitamins and neurological dysfunction. In: Morley JE, Sterman MB, Walsh JH (eds) Nutrional modulation of neural function. Academic Press, New York, pp 155–164 (UCLA forum in medical sciences, no. 28)

Fernstrom JD (1983) Role of precursor availability in control of monoamine biosynthesis in brain. Physiol Rev 63:484–546

Fernstrom JD, Larin F, Wurman RJ (1973) Correlation between brain tryptophan and plasma neutral amino acid levels following food consumption in rats. Life Sci 13:517–524

Fernstrom JD, Fernstrom MH, Grubb PE, Volk EA (1985) Absence of chronic effects of dietary protein content on brain tryptophan concentrations in rats. J Nutr 115:1337–1344

Foot M, Cruz TF, Clandinin MT (1982) Influence of dietary fat on the lipid composition of rat brain synaptosomal and microsomal membranes. Biochem J 208:631–640

Goodwin JS, Goodwin JM, Garry PJ (1983) Association between nutritional status and cognitive function in a healthy elderly population. JAMA 249:2917–2921

Gould RJ, Ginsberg BH (1984) Biochemistry and analysis of membrane phospholipids: applications to membrane receptors. In: Venter JC, Harrison LC (eds) Membranes, detergents, and receptor solubilization. Alan R. Liss, New York, pp 65–83

Greenwood CE, Craig REA (1987) Dietary influences on brain function: implications during periods of neuronal maturation. In: Rassin DK, Haber B, Drujan B (eds) Basic and clinical aspects of nutrition and brain development. Alan R. Liss, New York, pp 159–216 (Current topics in nutrition and disease, vol. 16)

Greenwood CE, Winocur G (1989) Learning and memory is impaired in rats fed a high saturated fat diet. Behav Neural Biol (in press)

Hrboticky N, Leiter LA, Anderson GH (1985) Effects of L-tryptophan on short-term food intake in lean men. Nutr Res 5:595–607

Leprohon-Greenwood CE, Anderson GH (1986) An overview of the mechanisms by which diet affects brain function. Food Technol 40:132–138,149

Li ETS, Anderson GH (1982) Self-selected meal composition, circadian rhythms and meal responses in plasma and brain tryptophan and 5-hydroxytryptamine in rats. J Nutr 112:2001–2010

Li ETS, Anderson GH (1984) 5-Hydroxytryptamine: a modulator of food composition but not quantity? Life Sci 34:2454–2460

Matheson DF, Oei R, Roots BI (1981) Effect of dietary lipid on the acyl group composition of glycerophospholipids of brain endothelial cells in the developing rat. J Neurochem 36:2073–2079

McGee CD, Greenwood CE (1989) Effects of dietary fatty acid composition on macronutrient selection and synaptosomal fatty acid composition. J Nutr 119:1561–1568

Milner JD, Wurtman RJ (1985) Tyrosine availability determines stimulus-evoked dopamine release from rat striatal slices. Neurosci Lett 59:215–220

Oski FA, Honig AS, Helu B, Howanitz P (1983) Effects of iron therapy on behavior performance in nonanemic iron-deficient infants. Pediatrics 71:877–880

Peters JC, Harper AE (1985) Adaptation of rats to diets containing different levels of protein: effects on food intake, plasma and brain amino acid concentrations and brain neurotransmitter metabolism. J Nutr 115:382–398

Pollitt E, Saco-Pollitt C, Leibel RL, Viteri FE (1986) Iron deficiency and behavioral development in infants and preschool children. Am J Clin Nutr 43:555–565

Rand WM, Windhan CT, Wyse BW, Young VR (eds) (1987) Food composition data: a user's perspective. The United Nations University, The Bath Press, Avon, UK

Rosner B (1986) Fundamentals of Biostatistics. Duxbury Press, Boston, pp 264–265

Ryan-Harshman M, Leiter LA, Anderson GH (1987) Phenylalanine and aspartame fail to alter feeding behavior, mood and arousal in men. Physiol Behav 39:247–253

Sandstead HH (1986) Nutrition and brain function: trace elements. Nutr Rev 4[suppl]:37–41

Schwab EK, Conners CK (1986) Nutrient-behavior research with children: methods, considerations, and evaluation. J Am Diet Assoc 86:319–324

Shinitzky M (1984) Membrane fluidity and cellular functions. In: Shinitzky M (ed) Physiology of membrane fluidity, vol 1. CRC Press, Boca Raton, pp 1–51

Stubbs CD, Smith AD (1984) The modification of mammalian membrane polyunsaturated fatty acid composition in relation to membrane fluidity and function. Biochim Biophys Acta 779:89–137

Sved AF (1983) Precursor control of the function of monoaminergic neurons. In: Wurtman RJ, Wurtman JJ (eds) Nutrition and the brain, vol 6. Raven Press, New York, pp 224–275

Wurtman RJ, Wurtman JJ (1984) Nutrients, neurotransmitter synthesis and the control of food intake. In: Stunkard AJ, Stellar E (eds) Eating and its disorders. Raven Press, New York, pp 77–86

Wurtman RJ, Hefti F, Melamed E (1981) Precursor control of neurotransmitter synthesis. Pharmacol Rev 32:315–225

Youdim MBH, Green AR, Bloomfield MR, Mitchel BL, Heal D, Grahame-Smith DG (1980) The effect of iron deficiency on brain biogenic monoamine biochemistry and function. Neuropharmacology 19:259–267

Youdim MBH, Ben-Schachar D, Ashkenazi R, Yehuda S (1983) Brain iron and dopamine receptor function. In: Mandell P, DeFeudis FV (eds) CNS receptors: from molecular pharmacology to behavior. Raven Press, New York, pp 309–322 (Advances in biochemistry and psychopharmacology, vol 37)

Youdim MBH, Sills MA, Heydorn WE, Creed GJ, Jacobowitz DM (1986) Iron deficiency alters discrete proteins in rat caudate nucleus and nucleus accumbens. J Neurochem 47:794–799

Young SN (1986) The clinical psychopharmacology of tryptophan. In: Wurtman RJ, Wurtman JJ (eds) Nutrition and the brain, vol 7. Raven Press, New York, pp 49–88

Young SN, Tourjman SV, Teff KL, Pihl RO, Anderson GH (1988) The effect of lowering plasma tryptophan on food selection in normal males. Pharmacol Biochem Behav 31:149–152

Chapter 11

A Behavioral Scientist's Perspectives on the Study of Diet and Behavior

B. J. Rolls and M. Hetherington

Food intake and selection, the activators of good nutrition, are influenced by both physiologic and psychologic factors. Thus it is important that studies of nutritional practices should draw from the methodologies of the behavioral and physiologic sciences. A diversity of approaches to the study of diet and behavior have been discussed in this volume. Other chapters have described and evaluated these methods and have given numerous examples of how they can be put into practice. A specific task remaining is to discuss how studies of diet and behavior can benefit from the use of methods developed by the behavioral scientist. It is not possible to discuss all of the methods in the space available so a selective approach is necessary. A specific, timely issue illustrates the points raised, namely how food intake and selection are affected by various sweeteners.

The boundaries between the domain of the nutritionist and that of the behavioral scientist are not always clear. This is good because it indicates that a degree of cross-fertilization has already taken place between the disciplines. This discussion considers the psychologist's contribution to be the development of rigorous laboratory-based studies that are often concerned with experiential and hedonic aspects of eating behavior. Behavioral scientists have also contributed to our understanding of how particular psychologic characteristics of individuals can affect nutrition. Details of the methods used by psychologists to study eating and drinking behavior can be found in a recent book (Toates and Rowland 1987) as well as in some of the other chapters in this volume.

Animal Studies

While the focus of this chapter will be investigations in humans, I would like to make a few points about investigations in lower animals. Crnic (Chapter 5) has discussed some of the advantages of using lower animals to study the effects of nutrients: the environment and experience of the animals can be controlled and manipulated, studies of physiologic mechanisms are often more feasible, their development is more compressed, and there is perhaps less chance of bias entering into the interpretation of the behavior. Crnic cautions about extrapolating findings in animals to humans.

The question of the relevance of animal models to human behavior has recently become an issue while studying the effects of sweeteners such as saccharin or aspartame on behavior. Rats show an increased intake of laboratory chow following consumption of saccharin solution and show increased preference for foods associated with saccharin consumption (Tordoff 1988). In trying to understand this increase in food intake the authors have drawn from the methodologies of both the behavioral and physiologic sciences. They have considered hedonic, metabolic and osmotic explanations for the increase and in so doing show clearly that a multidisciplinary approach is necessary for the understanding of the complex problems of diet and behavior.

However, studies of the effects of sweeteners in rats raise the issue of the validity of the rat as a model for human behavior. Psychologists are concerned not only with whether substances will or will not be ingested but also with whether or not they are found to be palatable. For example, although rats will consume large quantities of saccharin solutions, it appears that concentrated sugar solutions are more palatable. That rats find saccharin solutions to be no more palatable than relatively dilute sugar solutions was indicated by the results of two-bottle preference tests (Sclafani and Nissenbaum 1985). Also, preference tests indicate that aspartame and cyclamate apparently do not taste sweet to the rat.

Two points arising from this example should be emphasized. First, the validity of an animal model cannot be taken for granted. In this case, it is important to question the relevance of a rat model for studies of the effects of sweeteners which may be influenced by the palatability of the substance. Clearly rats respond to some sweeteners in a manner different from humans. Second, it is often important to understand the hedonic response to foods or drinks because this can impact on metabolism. Palatability may differentially affect various physiologic changes which follow ingestion. For example, in a study in humans when two meals of the same composition but different palatabilities were compared, it was found that consumption of the more palatable meal led to greater increases in oxygen consumption, plasma insulin and sympathetic activity (LeBlanc and Brondel 1985). Psychologists have developed methods for assessing palatability and preference in both laboratory animals and humans which should be used in studies of diet and behavior.

Human Studies

Recently there have been a number of studies of the effects of sugars and aspartame on eating behavior in humans. Since these studies have sometimes yielded contradictory results, they provide a good opportunity to discuss some methodologic issues. Emphasis will be on short-term laboratory based experiments.

Laboratory Studies

It is in the laboratory that the highest control over extraneous variables can be exercised. Variables can be excluded, held constant or controlled. Through an understanding of the various hedonic, environmental (Schachter and Rodin 1974) and cognitive influences (Wooley 1972) on eating behavior, psychologists and nutritionists can work together to design studies which are not confounded by these influences.

In the laboratory, eating behavior is often divided into behavioral units. The meal can be regarded as one such unit which can be studied in terms of the factors that initiate, maintain and terminate eating. The experimenter can measure the amount of food eaten and can determine whether it correlates with physiologic changes or with subjective assessments of hunger before and after the meal.

Subjects can be selected to control for body weight, dietary habits, weight history, food preferences and demographics. Foods can be chosen to reduce the variety available in natural settings and can be chosen with specific questions in mind, such as the effects of energy density, palatability, intensity of flavor, or nutrient composition on food intake and selection. Questionnaires related to body weight and eating habits can be used to isolate individual traits which influence intake. Ratings scales can be administered to study subjective sensations such as appetite, hunger, satiety, mood, and pleasantness of foods before and after test meals.

Thus in the basic laboratory study of food intake, stimuli such as energy density or palatability are varied and responses such as changes in subjective ratings or amount and type of food selected are recorded. It has been suggested that since the subject is aware that food intake is being measured in the laboratory this may influence the behavior. Indeed many of the environmental stimuli are absent or altered in the laboratory relative to free-living conditions. Should this affect eating behavior, this would reduce the applicability of findings in the laboratory setting to more naturalistic conditions (external validity). At least one study has demonstrated the equivalence between food intake in the laboratory and food intake under free-living conditions (Obarzanek and Levitsky 1985).

If the determinants of human eating behavior are to be discovered, it is more scientifically stringent to observe, test and record behavior in an environment where greater control can be exerted and variables of interest can be systematically manipulated. We therefore advocate controlled laboratory studies defining critical variables before more naturalistic investigations are undertaken. Thus, community-based studies would often benefit from being preceded by laboratory-based studies which define critical variables which should be either controlled or measured.

In most studies of diet and behavior there is a trade-off between rigorous control and naturalistic situations. The design that is chosen should be determined by the question being asked. Ideally, a number of approaches should be tried with the hope that the laboratory-based and naturalistic studies will ultimately lead to similar conclusions.

Selection of Subjects

One of the most important decisions to be taken is that of who is to be tested. This is one of the main areas in which the psychologist can make a contribution to the study of diet and behavior.

Early studies of human eating behavior focused on finding differences between subjects of different body weight (i.e., normal weight vs. overweight) (Schachter and Rodin 1974). Subsequent studies of the psychologic determinants of ingestive behavior indicate that body weight alone does not provide a sound basis for dividing subjects into groups (see Spitzer and Rodin 1981). People in the same weight category may have very different histories and experiences and attitudes to food which can produce very different constraints on their eating behavior. For example, normal weight individuals may be naturally lean and therefore may have never been concerned about caloric regulation; on the other hand they may be individuals who have successfully controlled their body weight and who are still very aware of what they are eating and consciously restrain food intake.

Personality, learning, developmental and emotional factors can have important implications for dietary habits. Although there is still no general agreement on which of these factors should be considered when subjects are being categorized, there are instruments available which aid classification.

It has been shown that dieting or "restrained" subjects, whether overweight or normal weight, behave differently in the laboratory under certain conditions by eating more than unrestrained, non-dieting subjects (see Herman and Polivy 1980). Therefore it is necessary to evaluate the extent to which subjects display restraint or the tendency to restrict food intake to control body weight. Herman and Polivy (1980) have developed the Restraint Scale, a questionnaire which has successfully predicted eating behavior in controlled situations (Heatherton et al. 1988). However, since the Restraint Scale measures two factors – concern for dieting and weight fluctuation – it has been argued that the scale fails to distinguish between overweight individuals and purely restrained individuals who do not have a history of weight fluctuation (Drewnowski et al. 1982). Stunkard and Messick (1985) have devised the Eating Inventory which overcomes this problem to a certain extent. This questionnaire measures three factors: cognitive restraint (the extent to which an individual controls eating behavior through dieting), disinhibition (the extent to which an individual's eating can be disrupted or disinhibited by internal or external circumstances), and perceived hunger. The disinhibition factor is independent of weight change and predicts overeating in the laboratory by obese subjects (Shrager et al. 1983). In contrast to the Restraint Scale devised by Herman and Polivy, the cognitive restraint factor of the Eating Inventory has not yet been found to predict eating behavior in laboratory studies (Heatherton et al. 1988).

The extent to which each of these different factors influences eating behavior is still under investigation; however, the use of such questionnaires as the Restraint Scale and the Eating Inventory provides a useful means of subdividing subjects into groups according to eating habits.

Surprisingly, there is still no clear definition of the type of subjects that should comprise the normal control group in studies of diet and behavior. In our studies of normal controls we select for individuals who are not dieting and who score low on dietary restraint. However, particularly among females, one must question whether this really is a normal group, since many women diet and show preoccupation with food.

Although progress has been made towards understanding behavioral attributes that could affect eating behavior, there is still no consensus on the classification of subjects. Perhaps as more investigators start administering the Eating Inventory, guidelines for subject classification will become clearer.

In studies of eating behavior it is also important to screen subjects for attitudes towards eating, weight and body image, since abnormal attitudes can affect eating behavior and may reflect an underlying eating disorder. The Eating Disorders Inventory (Garner et al. 1983a) is an important instrument for identifying attitudes, beliefs and behaviors typically associated with disordered eating. Similarly the Eating Attitudes Test (Garner and Garfinkel 1979) is a valid index of symptoms associated with anorexia nervosa. Both of these instruments can be used to detect eating-disordered individuals (Garner et al. 1983b) or high-risk individuals suspected of engaging in bulimic or anorexic behaviors (Williams et al. 1986). Unless the experimenter is specifically interested in subjects with eating disorders, subjects who show a preoccupation with weight or eating should be excluded from eating studies.

Returning to the specific example of studies of sweeteners, some of the studies have not attempted to classify the subjects according to dietary restraint or eating attitudes, which may account in part for the conflicting findings. Body weight and dieting may affect the hedonic response to the sweet taste (see Spitzer and Rodin 1981). It seems clear that the presence of an eating disorder can affect the liking for sweetness (Drewnowski et al. 1987). Therefore a clear definition of subject populations could help to decrease the variance within and between studies of sweeteners.

Test Foods and Drinks

In choosing foods for studies of diet and behavior it is important to control not only for nutritional composition, but also for the way the food is perceived by the subject. Obviously to achieve the greatest control when investigating the effects of particular nutrients on behavior, the experimenter could offer pure nutrients. For example, in the studies of sweeteners, subjects have been offered solutions of the sweeteners in water with no added flavor (Cabanac 1971; Blundell and Hill 1986). This is a valid and highly controlled design, but the validity of extrapolating findings with such unfamiliar and probably unpalatable solutions to soft drinks must be questioned. Although some purity of experimental design may be lost, if the experimental question relates to consumption of foods rather than nutrients, familiar foods should be used.

Subjects in controlled eating studies should always be screened to ensure that they like and are prepared to eat the test foods to be used. Although this seems obvious, it is often not considered as a part of the routine subject selection process. We have found that using subjects who do not like the test foods can lead to meaningless results.

Many aspects of a test food other than its nutrient composition can affect the outcome of experiments. For example, previous pleasant and unpleasant experiences with foods will affect whether or not they are eaten (Rozin and Vollmecke 1986). Beliefs about the caloric content of foods (Wooley 1972) and previous experiences with consumption of foods (Booth et al. 1982) may also affect the amount eaten since individuals may learn to associate particular foods with a particular reduction of hunger. Specific aspects of food presentation such as whether it is visible (see Porikos et al. 1982) and how much is offered (Fuller 1980, unpublished PhD thesis, University of Birmingham, UK) can affect intake. Studies designed to explore the social, environmental and cognitive influences on eating behavior should be continued. Researchers should be aware of how sensitive eating behavior is to such psychologic variables and control for them in their investigations.

Measuring Food Intake and Eating Behavior

Various techniques to determine the rate and style of eating are available. If accurate continuous measurements of intake are to be made, direct measurement techniques are required. Some of these methods have been based on animal feeding techniques using food dispensing machines (Hashim and Van Itallie 1964). These instruments can be made available to dispense liquid or solid foods either continuously or at set meal times, and can dispense quantities of food at a rate determined by the investigator or *ad libitum* by the subject. I will briefly review some of the methods used to record food intake. A more detailed review can be found elsewhere (Hetherington and Rolls 1987).

Liquid food dispensers are based directly on dispensers used to feed laboratory animals (Jordan et al. 1966). A liquid diet can be taken from a calibrated reservoir which may be connected to a continuous recorder so that rate and amount consumed are determined. This method has been particularly useful for studies concerned with compensation following caloric dilution and for studies in which visual cues about the food are eliminated. A problem, however, is that it is not clear whether intake of liquids is controlled by thirst mechanisms or hunger mechanisms or some combination of both.

Silverstone et al. (1980) have devised a solid food dispenser based on the vending machine. This technique utilizes a commercially available snack dispenser which allows the subject to request a food unit by pressing a button which releases the food unit into a compartment from which the subject collects the food. The dispenser can be connected directly to a computer which keeps track of the rate of eating as well as calories and nutrients consumed. This technique has been utilized in a variety of studies to record intake and selection over long periods of time. It is clearly more similar to a natural eating situation than the liquid dispenser, but it still has limitations on the types of food that can be presented. Also, having to request food from a machine may affect intake.

Intake of a single food item can be continuously recorded by having subjects consume foods from a bowl placed on a concealed electronic balance connected directly to a computer (Kissileff et al. 1980, 1982a,b). The theoretical basis for the methods in which cumulative intake is recorded is that the rate of eating relates to hunger and satiety. The rate of eating during the initial part of a meal is thought to relate to appetite and the palatability of the food whereas the change in rate over time is thought to reflect satiety. It has been found, for example, that overweight subjects (Pudel and Oetting 1977), bulimic subjects (Kissileff et al. 1986), and the elderly (Meyer and Pudel 1983) do not show the normal deceleration of eating as a meal progresses, indicating that they have impaired satiety. This method requires more basic studies confirming the validity of the assumptions about the relationship between eating rate and hunger and satiety. If they support current hypotheses, recording eating rate could provide a useful tool to detect differences, not only between individuals but also between different foods and nutrients.

A number of methods have been devised to study the microstructure of eating, such as the number of bites and chews and pauses between bites as well as the relationship between eating and drinking in a meal. These include a collar with a balloon sensitive to swallowing used in conjunction with a strain gauge in a head set for measuring chewing (Bellisle and Le Magnen 1980), a telemetric spoon or fork which records every time the utensil enters the mouth (Moon 1979), and an oral sensor consisting of a strain gauge attached to a dental retainer which telemetrically records chewing and swallowing (Stellar and Shrager 1985). Less intrusive methods include covertly observing eating behavior or using video cameras to record the behavior (Hill and McCutcheon 1975, 1984). These various methods have been developed because it is thought that eating style might relate to amount consumed. If this were the case, it would offer the potential to have some control over food intake through alterations in eating style. As yet, clear eating styles related to amount consumed or body weight have not been defined.

Thus there are a number of laboratory methods for measuring food intake. It is possible that some of these methods, such as telemetric monitoring of food intake, will be useful in more naturalistic settings so that subjects will not have to keep their own diet records. As more investigators continuously record food intake or observe eating behavior instead of simply weighing the amount consumed, they may find that different foods or nutrients have readily distinguishable effects on eating behavior.

Subjective Assessments

In studies of diet and behavior, one of the advantages of using human subjects is that they can be asked about habitual food intake, preferences and selection. In addition, under specific conditions subjects can be asked to assess internal and external sensations before and after test meals. During this time, subjective assessments of hedonic responses to foods, mood, appetite, hunger, satiety, and somatic and gastric sensations can be monitored.

There are several techniques which can assess subjective experience. However, finding methods of measurement which are both valid and reliable is problematic, since by the very nature of subjectivity, responses are difficult to quantify, interpret and compare between subjects. It is particularly important to remember that subjective responses are highly individual and subjective ratings made by one subject are not necessarily equivalent to ratings made by another subject. Similarly, a subject's experience of a sensation or perception of a stimulus on one occasion is not necessarily equivalent to the same subject's response on another occasion. Thus interpretation of responses is highly dependent upon the experimental procedure and the subject population.

Early investigators employed nominal scales to assess hedonics. These scales only require the subject to indicate whether a substance is liked, disliked, or neutral. This is not a very powerful way of distinguishing between substances. Subjective ratings made on numeric scales can be analyzed to investigate changes in direction and magnitude over time within that individual and, with less power, between individuals. A number of methods including rank ordering, fixed point scales, visual analog scales, and multidimensional scaling have been used. The advantages and disadvantages of these various methods have been discussed elsewhere (Drewnowski and Moskowitz 1985; Hetherington and Rolls 1987). The method which appears to be most sensitive in adults is the visual analog scale.

It cannot be assumed that subjective ratings will correlate with, or be predictive of, food intake or preferences. In some studies little relationship has been found between food preference (the expressed degree of liking or disliking for a food when obtained in response to a food name) and acceptance (the expressed degree of liking or disliking for a food when obtained in response to a prepared sample of the food) (Cardello and Maller 1982). However, it has been argued that attitudes to food can be good predictors of food consumption (Shepherd and Stockley 1985). In order to understand the significance of subjective ratings, the relationship to actual ingestive behavior or physiologic correlates should be determined.

There is no standard questionnaire which is being used to assess subjective sensations related to hunger, satiety, mood, etc. At present each investigator asks questions thought to be important, but there is a clear need for more work to standardize this type of assessment. The response will depend on the specificity of the question, so it has been suggested that the more specific are questions relating to hunger or taste, the more reliable the response (Spitzer and Rodin 1981).

Some lessons can be learned from the studies of the effects of sweeteners on appetite. Blundell and Hill (1986) reported that aspartame can cause a paradoxical increase in appetite based just on subjective ratings. They found that preloads of glucose caused a decrease in rated motivation to eat and an increase in stomach fullness, whereas a solution of aspartame increased self-rated motivation to eat and decreased ratings of fullness. They did not, however, measure food intake in this study. We have found that ratings of hunger or changes in hunger following a preload sometimes relate to subsequent intake (Rolls et al. 1988a), but they do not invariably do so. Recently three separate groups have failed to find that aspartame increases appetite or actual food intake (Birch 1987; Anderson and Leiter 1988; Rolls et al. 1988b,c).

The changing hedonic responses to sweeteners have also been assessed. Most individuals find sweet tastes highly palatable, but the response to sweetness will not always be the same and may depend upon their state of repletion, or how recently they have consumed sweet foods. In 1971 Cabanac showed clearly that the pleasantness of the sweet taste changed over time as subjects consumed glucose solutions. It is not clear whether this changing hedonic response depends on the decrease in the body's need for sugar or whether it depends upon the sensory stimulation accompanying ingestion. For this reason it has been of interest to examine the effects of aspartame on the pleasantness of the sweet taste.

Blundell and Hill (1986) compared the effects of glucose and aspartame solutions on the pleasantness of 20% sucrose solutions. They found that aspartame was about half as potent as a glucose load in decreasing the pleasantness of the sucrose solution over the hour after the loads. However, it is not clear whether hedonic ratings of solutions of sugar or sweeteners have any relevance to normal ingestive behavior.

The authors have developed a standard procedure using regular foods or drinks to assess hedonic responses for their sensory qualities (Rolls et al. 1981, 1985).

After documenting that rated hunger is similar on different test days, we offer subjects a tray containing small samples of eight or nine different foods or drinks. They are then instructed to taste and rate the pleasantness of the taste at that time on a visual analog scale. The samples are always offered in the same order and subjects rate each one in turn. The foods on the tray will include any item or items to be included in a test meal which is consumed just after assessments are finished. Test trays with the same foods can be offered over time after the end of the meal and changes in ratings both within foods and between foods can be analyzed. By looking at changes in ratings rather than absolute ratings, and by running subjects as their own controls in different test conditons, potentially confounding variables such as differences in initial pleasantness or individual preferences for foods are controlled.

Using this method we have determined the effect of eating foods sweetened with either sucrose or aspartame on the changing palatability of foods, or "sensory-specific satiety". We have found that desserts sweetened with aspartame are just as effective as those sweetened with sucrose in decreasing the pleasantness of the food that was consumed (Rolls et al. 1985, 1988b,c). We have found in other studies (Rolls et al. 1984) that these changes in pleasantness are related to subsequent consumption of foods.

Our studies of sensory-specific satiety can be used to illustrate how different disciplines can interact to understand issues of diet and behavior. The original impetus for our studies came from three different types of investigations. We knew from the early work of Davis (1928) that infants had a tendency to eat a variety of different foods when given choices. We also knew from the work of Cabanac (1971) that the hedonic response to particular tastes (i.e., sweet solutions) is not constant but decreases with consumption. Neurophysiologic studies in monkeys indicated that neurons in the hypothalamus responded to foods at the beginning of consumption when they were readily accepted, but as the animals became satiated the neurons became less responsive. However, the introduction of a different food reinstated both food acceptance and neuronal activity (Rolls and Rolls 1982). Our studies of sensory-specific satiety showed that the palatability of ordinary foods decreases during consumption, that this decrease in palatability occurs primarily for the food ingested or very similar foods, and that the changing response depends to a great extent on the sensory stimulation accompanying ingestion (Rolls 1986). This finding, which developed primarily out of behavioral and physiologic observations, has important implications for nutrition. Because satiety is relatively specific to a particular food that has been consumed, a variety of foods will be consumed to maintain palatability at a high level. Thus sensory-specific satiety helps to maintain a balanced diet because it encourages consumption of a variety of foods.

Conclusions

As the nutritionist discovers new ways in which nutrients affect behavior, metabolism, etc., there will be a need to understand how to implement dietary changes by altering food choices. Here the behavioral scientists, with their concern for issues of palatability and food acceptance and how individual characteristics influence food selection, may be able to help. Although the behavioral scientist has not yet developed reliable methods to change eating habits permanently, the advances that are being made make this more likely.

The issues related to diet and behavior are so complex that there is a need for all of the related disciplines to combine forces and use a variety of approaches. Definitive answers are elusive, so complementary data from different experimental approaches is the best hope for gaining understanding. As different sciences continue to grow both in the total knowledge base and in the sophistication of methodologies, there will be a greater and greater need for interdisciplinary collaborative studies of diet and behavior.

References

Anderson GH, Leiter L (1988) Effects of aspartame and phenylalanine on meal-time food intake of humans. Appetite 11[suppl]:48–53

Bellisle F, LeMagnen J (1980) The analysis of human feeding patterns: the edogram. Appetite 1:141–150

Birch LL (1987) The quantity and quality of young children's *ad libitum* consumption is influenced by the caloric density and type of sweetener in preloads. Fed Proc 46:1340–1341

Blundell JE, Hill AJ (1986) Paradoxical effects of an intense sweetener (aspartame) on appetite. Lancet i:1092–1093

Booth DA, Mather P, Fuller J (1982) Starch content of ordinary foods associatively conditions human appetite and satiation indexed by intake and eating pleasantness of starch-paired flavours. Appetite 3:163–184

Cabanac M (1971) Physiological role of pleasure. Science 173:1103–1107

Cardello AV, Maller O (1982) Relationships between food preferences and food acceptance ratings. J Food Sci 47:1553–1557

Davis CM (1928) Self selection of diet by newly weaned infants. Am J Dis Child 36:651–679

Drewnowski A, Riskey D, Desor JA (1982) Measures of restraint: separating dieting from overweight. Appetite 3:282–283

Drewnowski A, Moskowitz HR (1985) Sensory characteristics of foods: new evaluation techniques. Am J Clin Nutr 42:924–931

Drewnowski A, Bellisle F, Aimez P, Remy B (1987) Taste and bulimia. Physiol Behav 41:621–626

Garner DM, Garfinkel PE (1979) The Eating Attitudes Test: an index of the symptoms of anorexia nervosa. Psychol Med 9:273–280

Garner DM, Olmstead MP, Polivy J (1983a) The Eating Disorder Inventory: a measure of the cognitive behavioral dimensions of anorexia nervosa and bulimia. In: Darby PL, Garfinkel PE, Garner DM, Coscina DV (eds) Anorexia nervosa: recent developments. Alan R. Liss, New York

Garner DM, Olmstead MP, Polivy J (1983b) Development and validation of a multidimensional Eating Disorders Inventory for anorexia nervosa and bulimia. Int J Eat Dis 2:15–34

Hashim SA, Van Itallie TB (1964) An automatically monitored food dispenser apparatus for the study of food intake in man. Fed Proc 23:82–87

Heatherton TF, Herman CP, Polivy J, King GA, McGree ST (1988) The (mis)measurement of restraint: an analysis of conceptual and psychometric issues. J Abnorm Psychol 97:19–28

Herman CP, Polivy J (1980) Restrained eating. In: Stunkard AJ (ed) Obesity. Saunders, Philadelphia, pp 208–225

Hetherington M, Rolls BJ (1987) Methods of investigating human eating behavior. In: Toates F, Rowland N (eds) Feeding and drinking. Elsevier, Amsterdam, pp 77–109

Hill SW, McCutcheon NB (1975) Eating responses of obese and non-obese humans during meals. Psychosom Med 37:395–401

Hill SW, McCutcheon NB (1984) Contributions of obesity, gender, hunger, food preference and body size to bite size, bite speed and rate of eating. Appetite 5:73–83

Jordan HA, Wieland WF, Zebley SP, Stellar E, Stunkard AJ (1966) Direct measurement of food intake in man: a method for the objective study of eating behavior. Psychosom Med 28:836–842

Kissileff HR, Klinsberg G, Van Itallie TB (1980) Universal eating monitor for continuous recording of solid or liquid consumption in man. Am J Physiol 238:R14–R22

Kissileff HR, Thornton J (1982a) Facilitation and inhibition in the cumulative food intake curve in man. In: Morrison AJ, Stick P (eds) Changing concepts of the nervous system. Academic Press, New York, pp 585–603

Kissileff HR, Thornton J, Becker E (1982b) A quadratic equation adequately describes the cumulative food intake curve in man. Appetite 3:255–272

Kissileff HR, Walsh BT, Kral JG, Cassidy SM (1986) Laboratory studies of eating behavior in women with bulimia. Physiol Behav 38:563–570

LeBlanc J, Brondel L (1985) Role of palatability on meal-induced thermogenesis in human subjects. Am J Physiol 248:E333–E336

Meyer JE, Pudel V (1983) Das Essensverhalten im Alter und seine Konsequenzen für die Ernährung. Z Gerontol 16:241–247

Moon RD (1979) Monitoring human eating patterns during the ingestion of non-liquid foods. Int J Obes 3:281–288

Obarzanek E, Levitsky DA (1985) Eating in the laboratory: is it representative? Am J Clin Nutr 42:323–328

Porikos KP, Hesser MF, Van Itallie TB (1982) Caloric regulation in normal weight men maintained on a palatable diet of conventional foods. Physiol Behav 29:293–300

Pudel VE, Oetting M (1977) Eating in the laboratory: behavioral aspects of the positive energy balance. Int J Obes 1:369–386

Rolls BJ (1986) Sensory-specific satiety. Nutr Rev 44:93–101

Rolls BJ, Rolls ET, Rowe EA, Sweeney K (1981) Sensory specific satiety in man. Physiol Behav 27:137–142

Rolls BJ, van Duijvenvoorde PM, Rolls ET (1984) Pleasantness changes and food intake in a varied four-course meal. Appetite 5:337–348

Rolls BJ, Hetherington M, Burley BJ, van Duijvenvoorde PM (1985) Changing hedonic responses to foods during and after a meal. In: Kare MR, Brand JG (eds) Interaction of the chemical senses with nutrition. Academic Press, New York, pp 247–268

Rolls BJ, Hetherington M, Burley VJ (1988a) The specificity of satiety: the influence of foods of different macronutrient content on the development of satiety. Physiol Behav 43:145–153

Rolls BJ, Hetherington M, Laster LJ (1988b) Comparison of the effects of aspartame and sucrose on appetite and food intake. Appetite 11[suppl]:62–67

Rolls BJ, Hetherington M, Burley VJ (1988c) Sensory stimulation and energy density in the development of satiety. Physiol Behav 44:727–738

Rolls ET, Rolls BJ (1982) Brain mechanisms involved in feeding. In: Barker LM (ed) Psychobiology of human food selection. AVI, Westport, pp 33–62

Rozin P, Vollmecke TA (1986) Food likes and dislikes. Annu Rev Nutr 6:433–456

Schachter S, Rodin J (1974) Obese humans and rats. Lawrence Erlbaum, Halsted, Washington, DC

Sclafani A, Nissenbaum JW (1985) On the role of the mouth and gut in the control of saccharin and sugar intake: a re-examination of the sham-feeding preparation. Brain Res Bull 14:569–576

Shepherd R, Stockley L (1985) Fat consumption and attitudes towards food with a high fat content. Hum Nutr Appl Nutr 39A:431–442

Shrager EE, Wadden TA, Miller D, Stunkard AJ, Stellar E (1983) Compensatory intra-meal responses of obese women to reduction in the size of food units. Abst Soc Neurosci 628 (abstr)

Silverstone T, Fincham J, Brydon J (1980) A new technique for the continuous measurement of food intake in man. Am J Clin Nutr 33:1852–1855

Spitzer L, Rodin J (1981) Human eating behavior: a critical review of studies in normal weight and overweight individuals. Appetite 2:293–329

Stellar E, Shrager EE (1985) Chews and swallows and the microstructure of eating. Am J Clin Nutr 42:973–982

Stunkard AJ, Messick S (1985) The three-factor eating questionnaire to measure dietary restraint, disinhibition and hunger. J Psychosom Res 29:71–83

Toates FM, Rowland NE (eds) (1987) Feeding and drinking. Elsevier, Amsterdam

Tordoff MG (1988) Sweeteners and appetite. In: Williams GM (ed) Sweeteners: health effects. Princeton Scientific, Princeton, pp 53–60

Williams RL, Schaefer CA, Shisslak CM, Gronwaldt VH, Comerci GD (1986) Eating attitudes and behaviors in adolescent women: discrimination of normals, dieters, and suspected bulimics using the Eating Attitudes Test and Eating Disorders Inventory. Int J Eat Dis 5:879–894

Wooley SC (1972) Physiologic versus cognitive factors in short term food regulation in the obese and non-obese. Psychosom Med 34:62–68

Discussion: Implications for Future Research

M. R. C. Greenwood

Looking to the Future

Looking into the future is always fun, frustrating, and frequently fantasy. We can see the obvious directions our work and the work of others will take, but there is infrequently a real possibility of describing the newly synthesized ideas that prove to be true insight. Nonetheless, there are few things that intrigue scientists more than the study of behavior. Even among "hard scientists", the behavior of particles and subatomic particles, not to mention the behavior of intergalactic spaces, remain topics rife with controversy which frequently spawn new, and even amusing, scientific concepts and vocabulary.

Many scientists who try to study diet and behavior feel a certain frustration that quantitation and causal relations are not easily sorted out. This is to be expected when multiple and variable processes are studied conjointly. In a biological age when much excitement centers on dissecting individual genes, there is no need to apologize about the urge to quantify complex mechanisms whose outcome variables may be the composite of much more intricate processes and about which we know very little. The way an individual scientist responds to scientific methodology is conditioned by his/her individual acculturation to science and to the disciplines or subdisciplines which are best understood. Scholars of integrative process recognize that great insights are not always dependent upon, or tied to the development of, reductionist tools, although new scientific insights are frequently aided by such tools. The fact that investigators studying diet and behavior may need to coordinate and use the reductionist tools already available as well as be ready to exploit new ones has been amply demonstrated in this volume.

For example, each of the contributors clearly understands both the power and the limitations of his/her own research and his/her own research subspecialty. Each has clearly demonstrated how difficult it is to design the perfect experiment. In turn, the communication between investigators primarily interested in diet and those primarily interested in behavior has been a provocative illustration.

I believe we have made an excellent start at defining and acknowledging the multiple tools and types of approaches which are necessary for us to increase our understanding of the relationships between diet and behavior. However, we have a good deal of work left to do on advancing the development of research methods and standards and in encouraging multidisciplinary interaction among researchers in the field. This is not intended as a criticism but rather as a realization of the fact that the development of research methodology in diet and behavior is still incomplete.

The Challenges and Problems of Multidisciplinary Interactions

The understanding of behavior, how it influences variables such as diet and vice versa, is an integral component of the frontier of the next biological revolution. All, or nearly all, of the major social problems of our society are related to, and are contingent upon, our ability to grasp the fundamentals of human behavior and to understand and affect those behaviors. We are in the middle of an extraordinary explosion of scientific information brought on, at least in part, by the development of the new molecular biology tools. The potential of these tools to aid us in our exploration of human behavior is still largely unexploited.

Within a few years, and most certainly within a decade, 1000, or perhaps considerably more than 1000, genetic markers will be available and readily testable so that linkage analysis of human phenomena, such as specific behaviors, can proceed. The areas of molecular biology that will allow us to localize traits on certain human chromosomes and to provide flanking markers are already generating new knowledge and excitement. We are cautioned (see Ward, Chapter 9), and rightly so, that although the use of molecular biology tools is a completely satisfactory approach, and molecular biology will provide genetic markers, if we want to study a behavior, its genetic, nutritional, or its environmental modulation, we must be able to specify this behavior and readily quantitate it carefully and reproducibly. This is also true for quantitating diet. For example, if we do find a gene that is responsive to a high fat diet, this will not be helpful in studying human behavior if we cannot say with reliability whether, or in what amounts, individuals carrying the gene ingest the dietary stimulus.

We must also realize that knowing the probable chromosomal localization or locations of a genetic trait still cannot tell us much about the etiologic factors leading to the expression of this trait. In fact the "new genetics", which holds great promise, can also lead to interesting and somewhat bizarre twists in the way scientists approach biological phenomena. For most scientists who learned to study biochemistry or cell biology 10 years ago or longer, one of the traditional approaches, and one still used very frequently, is to determine a function, to identify the enzyme or the pathway which regulates this function, and to then try to understand the cellular and molecular control of this pathway.

Now, with the powerful tools of the new molecular biology, bits of the genome, in a particular location on particular chromosomes, can be cloned. They can be expressed in vectors, and then their protein products evaluated. It is very interesting that now, instead of functions looking for a regulatory protein, we have proteins looking for a function. Consequently, molecular biologists are rapidly discovering that they need the traditional biological specialists and the behavioral scientists in order to develop their functional hypotheses. Of course, the behavioral scientists and the nutritional scientists also need to recognize the power that some of these new tools offer and to become sufficiently literate in their use to participate in the interpretation of the generative data.

Clearly, the implications of the molecular and genetic approaches for future research are promising. Nonetheless, in order to exploit the new opportunities and knowledge, the precision with which we measure and assess diet, behavior and the environment must be improved. Much has been said about this, and yet the tools that are needed to move this area of research forward seem to be emerging very slowly.

The Needs for New Methods in Diet and Behavior

New methods and standards will need to be developed and multidisciplinary interactions encouraged. One of the things that may help to subdivide hetero-genous human populations, particularly when the interest is in food interactions, is the genomic marker that allows us to distinguish one group of individuals from another. In addition to genomic markers, methods need developing which allow the decrease of heterogeneity and better subject selection. In the area of energy balance, some recent work has been helpful in predicting a biobehavioral outcome, i.e., weight gain. For example, an intriguing study was reported by Roberts et al. (1988) in which a measure of adiposity (BMI) was determined in children that were born to normal weight mothers and to obese mothers. Among those children born to normal weight mothers, a steady weight gain over 18 months occurred. Among children born to obese mothers, two groups were noted. One group had a BMI identical to those of children born of non-obese mothers, but another group had BMIs indicating increased adiposity. These outcome observations were particularly interesting because the study had previously measured the total energy expenditure for the infants when they were less than 3 months old. Children born to normal weight mothers and some infants born to obese mothers had a normal and identical resting energy expenditure, whereas the children born to overweight mothers who subsequently had increased BMIs themselves had a lower energy expenditure than the infants born to normal weight mothers.

These data suggest that there may be a functionally measurable component of energy, which does not necessarily await new genetic tools, that would allow us to distinguish between certain categories of individuals, such as those who may be less responsive to energy expenditure. Furthermore, this functional, predictive component may also be useful in adults. For example, Ravussin et al. (1988) studying the Pima Indians, have been able to predict, using metabolic rates, which individuals would gain 10 kilograms within a period of 1 to 3 years.

The Need for Additional Animal Models

Another aspect of research that it is very important to continue to develop is the elucidation of better environmentally induced and genetic animal models of disease. There are many types of examples of animal models that might be further developed. Several points may be illustrated by using models of genetic obesity (Bray and York 1971). Many of the mouse models exist as inbred strains and the chromosomal location of the gene conveying the adiposity is known. However, mice are very difficult to use when studying feeding behavior and/or diet selection and, consequently, much use has been made of genetically obese rat models. Perhaps, the most popular of these models is the genetically obese Zucker (*fafa*) rat. This rat inherits the obesity in a Mendelian recessive fashion. Furthermore, the rat becomes obese even in the absence of hyperphagia (Cleary et al. 1980), and remains obese even in the face of pharmacologic (Greenwood et al. 1981a; Vasselli et al. 1983), surgical (Greenwood et al. 1982) or exercise (Walberg et al. 1983) intervention. The author's laboratory has studied the early developmental course of the development of the *fafa* obesity and has proposed that early alterations in the regulation of the "gatekeeper" enzyme lipoprotein lipase (LPL) are essential to the development of the obesity (Greenwood et al. 1981b), although the genomic basis of the obesity is still under study. A new variant on the genetically obese *fafa* rat has recently been described (Ikeda et al. 1981). This new Wistar fatty (*fafa*) is an albino Wistar rat with the *fa* gene. In this model, the obesity occurs in both males and females. Like the Zucker *fafas* the females do not become diabetic. However, unlike the Zucker *fafas*, the males become frankly diabetic (Kava et al. 1990). Thus, the same gene (*fa*) can lead to differing associated pathologies in two unrelated background strains and both homozygous and heterozygous effects can be studied. The use of such models in understanding the relationship between diet and disease should enhance our ability to dissect genetic and environmental effects.

In addition to the genetic models, there is a wide variety of animal models where the obesity is produced by diet manipulation. A recent model of interest allowed female rats to select their own diet from three dietary sources. When the rats were allowed to do so, they selected a diet containing over 35% fat and became obese compared with chow fed controls. Perhaps even more impressive was the observation that switching the rats back and forth between self-selection and chow resulted in even more obese rats, with the adiposity prominent in the abdominal fat depots (Reed et al. 1988). This increase in abdominal fat pad weight occurred without marked "weight cycling" and was accompanied by increased plasma insulin levels (Reed et al. 1988). Thus, new dietary models may also allow us to understand how dietary behavior leads to changed metabolism and enhanced risk for diseases such as hypertension and diabetes. There are many more such examples, and there is a need for more investigators to familiarize themselves with the strengths and limitations of the available animal models to study the interactions of diet and behavior.

New Models in Human Behavior

In addition to new animal models that allow an understanding of the association between genetic and diet interactions, better behavioral measures for assessing human behavior are also needed. For example, Drewnowski et al. (1985) have

developed simple psychophysical evaluations which distinguish hedonic preferences between individuals who have either disordered feeding behaviors or
disordered metabolism.

This simple test, in which patients are asked to test 20 solutions and rate them
both on magnitude estimate and on a hedonic scale, may potentially be a useful
screening test for taste preferences in some human populations. For example,
normal weight individuals respond with a typical two-dimensional grid showing a
normal individual's sweetness and fat preferences of approximately 8% sugar and
20% fat. Obese individuals have a lower preference for sugar but a higher
preference for fat, while reduced obese individuals have even higher fat
preferences and in addition their sugar preference is returned toward normal
(Drewnowski et al. 1987). Interestingly, a comparison between bulimic individuals and control subjects demonstrates the lack of the fat preference. Bulimics do
show a sweet preference consistent with their reported carbohydrate phobia.

Another important approach to quantitating human feeding behavior may be
the oral sensor. The one being developed at University of Pennsylvania is
potentially a useful field tool because it can be worn for considerable periods of
time, and it monitors chewing behavior. The oral sensor cannot tell you how
much an individual has eaten, but it can be used to assess frequency of ingestion
and speed of ingestion. Much additional work remains before these newer
devices can be routinely used in free-feeding situations, but they promise to be
useful in the future. When it is considered that only a 2% error in caloric intake
estimates makes it impossible to predict substantial changes in body weight over a
period of a year or more, then there are serious challenges in validating methods
that will allow us to quantitate ingestion in humans. We need to encourage
serious collaboration with social scientists who are experts in survey analyses, and
with cognitive psychologists who are experts in the methods of training and
behaviors, and who understand how individuals report information. We must
also find new ways to harness the powerful microcomputer technology and use it
to monitor food consumption. Better measures of cultural and environmental
input are also needed.

There is a great need to develop collaborative mechanisms which allow true
cross-disciplinary collaboration between members of the neurosciences, behavioral science and the nutritional communities, as well as those which increase
contact with the molecular and cellular biologists. In addition, we must establish
more effective communication networks. This is always very difficult. Most
communication problems derive in part because of the nature of collaborative
endeavors. There are two primary types of collaboration. The predominant form
of much scientific collaboration is of the "opportunistic" sort, i.e., the addition of
new measures to a study originally designed for another purpose. In such a case,
each investigator uses his/her expertise, and the results are analyzed according to
subdisciplinary areas.

The other type of collaboration is more difficult. It requires more time and
effort than many are willing to invest. This form of collaboration requires a group
of individuals who are interested in the same problem to design studies together
and to derive experiments which are different from those each would have done
individually. An example of such collaboration began for the author's laboratory
in 1984 when we became involved in the MacArthur Foundation sponsored
project: the "Weight Cycling Project." Our collaborative group shared an
interest in human behavior and metabolism associated with repeated episodes of

weight loss and regain. Little is known about the long-term impact of repeated episodes of weight loss and regain behaviorally, nutritionally or on long-term health. This is the type of interdisciplinary problem that can be difficult to get funded through the usual government agencies because it requires epidemiologic work, animal experimental work and clinical work, and it is necessary to bring together individuals who do not normally work together. For example, our collaborative group includes surgeons, dietitians, exercise physiologists, cell and molecular biologists, and nutrition scientists.

This collaborative group required of all core participants a commitment to spend 3 days at least twice a year together, talking about experimental design and interpretation. It has taken a great deal of effort but we are encouraged to think that the type of scholarship that we have begun to develop is reflective of a truly collaborative input. There is still a considerable period of time on the project ahead of us before we know the overall outcome, but this collaboration has changed the research approaches of several independent groups. Perhaps the diet and behavior field would benefit from more collaborative groups of this sort. It may be useful to persuade private or public agencies that this is a worthwhile expenditure even though a portion of the funding is spent on meetings and collaborative brainstorming instead of on experimental protocols *per se*.

Finally, we must think about training the next generation of scientists and interesting them in issues related to diet and behavior. This is not going to be an insignificant challenge because there is great pressure on students at the undergraduate level, even at school level, to narrow the focus of their education. This is not confined to science education, but affects their entire education. We must increase the probability that the next generation of scientists will be able to incorporate and integrate information from different disciplines.

If we are not successful in our efforts to teach integrative thinking, we will not have scientists who are interested in studying multidisciplinary areas such as diet and behavior. We need to inculcate into the training of scientists studying chemistry, physics, psychology or social science, an appreciation for the broad social issues, such as the impact of what we eat on how we feel, how we perform and how society is affected by this interaction. These questions need to be made relevant in our undergraduate and graduate educational programs so that even when individuals specialize in molecular genetics, physiologic psychology or cultural anthropology, they will find it appropriate to ask about and incorporate broader areas of knowledge.

Summary

A great deal of progress has been made in individual subdisciplines related to diet and behavioral interactions. However, in order for the field to advance rapidly over the next decade it is necessary to put together collaborative groups representing the best talents in various subdisciplines. Such groups are hard to identify and to support within much of the current government and private research support systems. Thus, new initiatives will be necessary. In addition, consideration needs to be given to encouraging young investigators to enter the fields related to diet and behavior. In an age of increasing subspecialization, some new programs that encourage cross-disciplinary, integrative thought must emerge. In short, the relationships inherent between diet and behavior are among the most exciting and challenging for us to confront as we approach the next age of neuroscience investigation, but, in order to move forward, it will be necessary to develop both technical and human resources.

References

Bray GA, York DA (1971) Genetically transmitted obesity in rodents. Physiol Rev 51:598–646

Cleary MP, Vasselli JR, Greenwood MRC (1980) Development of obesity in the Zucker (*fafa*) rat in the absence of hyperphagia. Am J Physio 238:E284–E292

Drewnowski A, Brunzell JD, Snade K, Iverius PH, Greenwood MRC (1985) Sweet tooth reconsidered: taste responsiveness in human obesity. Physiol Behav 35:617–622

Drewnowski A, Halmi KA, Gibbs J, Pierce B, Smith GP (1987) Taste and eating disorders. Am J Clin Nutr 46:442–450

Greenwood MRC, Stern JS, Triscari J et al. (1981a) The effect of hydroxycitrate on adipose tissue development in the Zucker obese rat. Am J Physiol 240:E72–E78

Greenwood MRC, Cleary MP, Steingrimsdottir L, Vasselli JR (1981b) Adipose tissue metabolism and genetic obesity: the LPL hypothesis. Recent Adv Obesity Res III:75–79

Greenwood MRC, Maggio CA, Koopmans HS, Sclafani A (1982) Zucker *fafa* rats maintain their obese composition ten months after jejunoileal bypass surgery. Int J Obesity 6:513–525

Ikeda H, Shino A, Matsao T et al. (1981) A new genetically obese, hyperglycemic rat (Wistar fatty). Diabetes 30:1045–1050

Kava RK, West DB, Lukasik VA, Greenwood MRC (1990) Sexual dimorphism of hyperglycemia and glucose tolerance in the Wistar fatty rat, Diabetes (in press)

Ravussin E, Lillioja S, Knowler WC et al. (1988) Reduced rate of energy expenditure as a risk factor for body weight gain. N Engl J Med 318:467–472

Reed D, Contreras RJ, Maggio C, Greenwood MRC, Rodin J (1988) Weight cycling in female rats increases dietary fat selection and adiposity. Physiol Behav 42:389–395

Roberts S, Savage J, Coward WA et al. (1988) Energy expenditure and intake in infants born in lean and overweight mothers. N Engl J Med 318:461–466

Vasselli JR, Haracczkiewicz EH, Maggio CA, Greenwood MRC (1983) The effects of a glucosidase inhibitor (BAY g 5421) on the development of obesity and food motivated behavior in obese (*fafa*) rats. Pharmacol Biochem Behav 19:85–95

Walberg J, Greenwood MRC, Stern J (1983) Lipoprotein lipase activity and lipolysis following swim training in obese and lean Zucker rats. Am J Physiol 245:R706–712

Subject Index